Lecture Notes i
Computer Science

Edited by G. Goos and J. Hartmanis

25

Category Theory Applied to Computation and Control

Proceedings of the First International Symposium
San Francisco, February 25–26, 1974

Edited by E. G. Manes

Springer-Verlag
Berlin · Heidelberg · New York 1975

Dr. Ernest Gene Manes
The Commonwealth of Massachusetts
University of Massachusetts
Dept. of Mathematics and Statistics
Arnold House
Amherst, MA 01002/USA

Library of Congress Cataloging in Publication Data

Main entry under title:

Category theory applied to computation and control : proceedings of the first international symposium, San Francisco, Febr. 25-26, 1974.

(Lecture notes in computer science ; 25)
"An A. M. S. symposium within the annual meeting of the American Association for the Advancement of Science."
Bibliography: p.
Includes index.
1. Machine theory--Congresses. 2. Automata--Congresses. 3. Control theory--Congresses. 4. Categories (Mathematics)--Congresses. I. Manes, E. G., 1943- ed. II. American Mathematical Society. III. American Association for the Advancement of Science. IV. Series.
QA267.C35 629.8'312 74-34481

AMS Subject Classifications (1970): 00A10, 00A15, 02F05, 02F10, 02F15, 02F20, 02F29, 02F35, 06A20, 13C99, 18A40, 18B20, 18C15, 18D15, 22E99, 34A30, 34C35, 34H05, 60A05, 68-02, 68A25, 68A30, 68A40, 73B30, 93A05, 93A10, 93B05, 93B10, 93B15, 93B20, 93B25, 93B30, 93C15, 93C25, 93E99, 94A25, 94A30, 94A35
CR Subject Classifications (1974): 5.20, 5.21, 5.22, 5.23, 5.24

ISBN 3-540-07142-3 Springer-Verlag Berlin · Heidelberg · New York
ISBN 0-387-07142-3 Springer-Verlag New York · Heidelberg · Berlin

Offsetdruck: Julius Beltz, Hemsbach/Bergstr.

PREFACE

This book presents the results of a symposium which brought together scientists interested in applying modern algebraic techniques to problems in control and in computer science. In addition to extended abstracts of the contributed papers, we have included introductory material to provide the expert in control or computation with the necessary background in category theory (introduction, part 1); to present some of the problems of control theory still to be satisfactorily placed in a categorical framework (part 2); and to give those fluent with category theory an overview of its applications in computation and control (part 3). We also include a bibliography at the end of the volume.

The First International Symposium: Category Theory Applied to Computation and Control was arranged by E. G. Manes at the invitation of S. Mac Lane, President of the American Mathematical Society. The Program Committee comprised M. A. Arbib, E. S. Bainbridge, J. A. Goguen, E. G. Manes and M. Wand. As an A. M. S. Symposium within the Annual Meeting of the American Association for the Advancement of Science, the Symposium was held at the St. Francis Hotel in San Francisco on February 25 and 26, 1974.

A limited edition of these Proceedings was published by the Mathematics Department and the Department of Computer and Information Science of the University of Massachusetts at Amherst in February, 1974. The abstracts of Carlson, Ehrig-Kühnel-Pfender and Hoehnke did not appear in this edition.

The support of the National Science Foundation under Grant Number GJ 35759 --both in the preparation of the contributions by Alagić, Anderson, Arbib and Manes and in the publication of the limited edition-- is gratefully acknowledged.

E. G. Manes
Amherst, Massachusetts
September, 1974

TABLE OF CONTENTS

PROGRAM AND PARTICIPANTS

Categories; use and misuse, S. EILENBERG†, Columbia University, New York

Semantics of computation, J. A. GOGUEN†, University of Southern California at Los Angeles

Linear systems over rings of operators, B. F. WYMAN, Ohio State University

Duals of input/output maps, B. F. WYMAN (with J. RISSANEN, Linköping University)

Control of linear continuous-time systems defined over rings of operators, E. KAMEN, Georgia Institute of Technology

Representation of a class of nonlinear systems, E. T. ONAT, Yale University, (with J. A. GEARY, Yale University)

Addressed machines and duality, E. S. BAINBRIDGE, University of Ottawa

Time-varying systems, M. A. ARBIB, University of Massachusetts, (with E. G. MANES, University of Massachusetts)

Some structural properties of automata defined on groups, A. WILLSKY, Massachusetts Institute of Technology (with R. BROCKETT, Harvard University)

Automata in semimodule categories, J. MESEGUER, Universidad de Zaragoza (with I. SOLS, Universidad de Zaragoza)

Realization is continuously universal, L. A. CARLSON, University of Chicago

Power and initial automata in pseudoclosed categories, H. EHRIG, Technische Universität Berlin (with H.-J. KREOWSKI, Technische Universität Berlin)

Fuzzy morphisms in automata theory, E. G. MANES (with M. A. ARBIB)

Factorization of Scott-style automata, J. L. BAKER, University of British Columbia

Application of categorical algebra to classification of systems, M. MESOROVIC and Y. TAKAHARA, Case Western Reserve University; presented by P. C. KAINEN, Case Western Reserve University

Complexity as a general mathematical idea: subadditive functions on the Grothendieck ring of a cateogry, J. RHODES, University of California at Berkeley

An abstract machine theory for formal language parsers, D. BENSON, Washington State University

An algebraic formulation of the Chomsky hierarchy and the recursive specification of data types, M. WAND, Indiana University

The algebraic theory of recursive program schemes, J. THATCHER, IBM Thomas J. Watson Research Center (with R. M. BURSTALL, University of Edinburgh)

Scattering theory and non linear systems, J. HELTON, Dowling College (with W. HELTON, State University of New York at Stony Brook)

† Invited addresses.

Cellular automata with additive local transition, W. MERZENICH, Universität Dortmund

Diagram Characterization of μ-recursion and while loops, M. PFENDER, Technische Universität Berlin (with H. EHRIG and W. KÜHNEL, Technische Universität Berlin)

The tricotyledon theory of system design, A. W. WYMORE, University of Arizona

J. D. Aczel, Department of Applied Analysis and Computer Science, University of Waterloo, Ontario, CANADA.

Michael Arbib, Department of Computer and Information Science, University of Massachusetts, Amherst, Massachusetts 01002, USA.

John Backus, 91 St. Germain Avenue, San Francisco, California 94114, USA.

Stew Bainbridge, Department of Mathematics, University of Ottawa, Ottawa, Ontario, K1N 6N5, CANADA.

John Baker, Department of Computer Science, University of British Columbia, Vancouver, Saskatchewan, CANADA.

David Benson, Department of Computer Science, Washington State University, Pullman, Washington 99163, USA.

Don Boucher, Department of Computer Science, Washington State University, Pullman, Washington 99163, USA.

Lee Carlson, Committee on Information Sciences, University of Chicago, Chicago, Illinois 60637, USA.

Roger Chetwynd, Department of Computer Science, Washington State University, Pullman, Washington 99163, USA.

Paul Cull, Oregon State University, Corvallis, Oregon 97331, USA.

Jane Day, College of Notre Dame, Belmont, California 94002, USA.

Bradley Dickinson, Stanford University, Stanford, California, USA.

Hartmut Ehrig, Fachbereich 20 - Kybernetik, Technische Universität Berlin, 1 Berlin 12, Strasse des 17. Juni 135, GERMANY.

Sammy Eilenberg, Department of Mathematics, Columbia University, New York, New York, USA.

L. Forman, University of Chicago, Chicago, Illinois 60637, USA.

Bruce Fowler, I. A., 5831 Sunset Blvd., Hollywood, California, USA.

Wallace Givens, Argonne National Laboratory, 9700 South Cass Avenue, Argonne, Illinois 60439, USA.

Susanna Ginali, University of Chicago, Chicago, Illinois 60637, USA.

Joe Goguen, Computer Science Department, 3532 Boelter Hall, University of California, Los Angeles, California 90024, USA.

John Gray, Department of Mathematics, University of Illinois, Urbana, Illinois 61801, USA.

Joanne Helton, Dowling College, Oakdale, New York, USA.

Fred Hornbeck, Department of Psychology, San Diego State University, San Diego, California 92115, USA.

André Joyal, Université du Québec, Montréal, CANADA.

Edward Kamen, Department of Electrical Engineering, Georgia Institute of Technology, Atlanta, Georgia 30332, USA.

Paul Kainen, Case Western Reserve University, Cleveland, Ohio, USA.

Jacqueline Klasa, Department of Mathematics, University of Ottawa, Ottawa, Ontario, K1N 6N5, CANADA.

Pierre Leroux, Université du Québec, Montréal, CANADA.

Fred Linton, Department of Mathematics, Wesleyan University, Middletown, Connecticut 06467, USA.

Saunders Mac Lane, Department of Mathematics, University of Chicago, Chicago, Illinois 60637, USA.

T. S. E. Maibaum, Department of Applied Analysis and Computer Science, University of Waterloo, Waterloo, Ontario, N2L 3G1, CANADA.

Ernie Manes, Department of Mathematics, University of Massachusetts, Amherst, Massachusetts 01002, USA.

Wolfgang Merzenich, Informatik I, Universität Dortmund, 47 Dortmund, den August-Schmidt-Strasse, Dortmund, GERMANY.

José Meseguer, Departamento de Electridad y Electronica, Universidad de Zaragoza, Zaragoza, SPAIN.

Turan Onat, Department of Electrical Engineering, Yale University, New Haven, Connecticut 06520, USA.

Michael Pfender, Mathematik, Technische Universität Berlin, 1 Berlin 12, Strasse des 17. Juni 135, Berlin, GERMANY.

John Rhodes, Department of Mathematics, Evans Hall, University of California, Berkeley, California 94720, USA.

Jorma Rissanen, Ïnstitutionen för Systemteknik, Linköping University, S-581 83 Linköping, SWEDEN.

Peter Sawtelle, University of Missouri, Rolla, Missouri, USA.

Michael Shantz, California Institute of Technology, 286-80, Pasadena, California 91109, USA.

Al Shpuntoff, University of Illinois, Urbana, Illinois 61801, USA.

Ignacio Sols, Departamento de Geometricia, Universidad de Zaragoza, Zaragoza, SPAIN.

Eduardo Sontag, Center for Mathematical System Theory, University of Florida, Gainesville, Florida 32601, USA.

Arthur Stone, Simon Fraser University, Burnaby, British Columbia, CANADA.

Jim Thatcher, IBM Thomas J. Watson Research Center, P.O. Box 218, Yorktown Heights, New York 10598, USA.

Mitchell Wand, Computer Science Department, 101 Lindley Hall, Indiana University, Bloomington, Indiana 47401, USA.

Alan Willsky, Department of Electrical Engineering, Massachusetts Institute of Technology, Cambridge, Massachusetts 02139, USA.

W. Wonham, Department of Electrical Engineering, University of Toronto, Toronto, Ontario, CANADA.

Bostwick Wyman, Department of Mathematics, 231 W. 18th Avenue, Ohio State University, Columbus, Ohio 43210, USA.

Wayne Wymore, Department of Systems Engineering, University of Arizona, Tucson, Arizona 85721, USA.

Bob Yacobellis, Bell Laboratories, Naperville, Illinois 60540, USA.

L. A. Zadeh, Department of Electrical Engineering and Computer Science, University of California, Berkeley, California 94720, USA.

BASIC CONCEPTS OF CATEGORY THEORY APPLICABLE TO COMPUTATION AND CONTROL

M. A. Arbib
Computer and Information Science

E. G. Manes
Mathematics

University of Massachusetts
Amherst, MA 01002 U.S.A.

The reader of papers in this symposium volume will find all the necessary definitions --and statements of most of the appropriate theorems-- from category theory concisely presented in this chapter. These concepts will appear with full motivation, examples, proofs and exercises in our forthcoming book "The Categorical Imperative: Arrows, Structures, and Functors" (Academic Press, 1974).

Mac Lane's "Categories for the Working Mathematician" (Springer-Verlag, 1972), Herrlich and Strecker's "Category Theory" (Allyn and Bacon, 1973) and the bibliographies there may be consulted for a wealth of further material.

The dependency of sections is shown as follows:

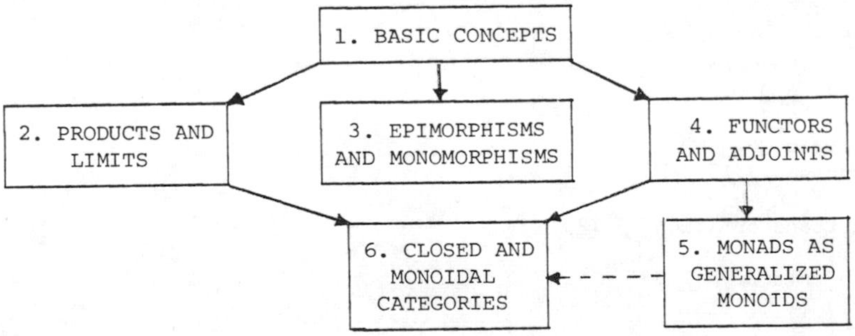

This will help the reader in deciding how far to back up in approaching the definition of a concept as located by the following index (a.b means definition b of section a):

1. BASIC CONCEPTS

1.1 DEFINITION: A category $\mathcal{K}$ comprises a collection $\mathrm{Obj}(\mathcal{K})$, called the set of objects of $\mathcal{K}$, together with for each pair A,B of objects of $\mathcal{K}$ a distinct set $\mathcal{K}(A,B)$ called the set of morphisms from A to B subject to the conditions I and II below. We may write $f : A \to B$ or $A \xrightarrow{f} B$ to indicate that the morphism f is in $\mathcal{K}(A,B)$, and we then refer to A as the domain of f and to B as the codomain of f:

I. For any three (not necessarily distinct) objects A, B and C of $\mathcal{K}$, there is defined a map

$$\mathcal{K}(A,B) \times \mathcal{K}(B,C) \to \mathcal{K}(A,C) \;:\; (A \xrightarrow{f} B,\ B \xrightarrow{g} C) \mapsto A \xrightarrow{g\cdot f} C$$

called composition which satisfies the associativity axiom that for all objects A,B,C,D of $\mathcal{K}$ and all morphisms f in $\mathcal{K}(A,B)$, g in $\mathcal{K}(B,C)$ and h in $\mathcal{K}(C,D)$ we have $h\cdot(g\cdot f) = (h\cdot g)\cdot f : A \to D$. We may write $A \xrightarrow{f} B \xrightarrow{g} C$ for $A \xrightarrow{g\cdot f} C$; the associativity axiom then says that we may write longer arrow-chains such as $A \xrightarrow{f} B \xrightarrow{g} C \xrightarrow{h} D$ without ambiguity as to the composite morphism $h\cdot g\cdot f = (h\cdot g)\cdot f = h\cdot(g\cdot f)$ which is indicated.

II. For every object A of $\mathcal{K}$, the set $\mathcal{K}(A,A)$ contains a morphism id_A, called the identity of A, with the property that for every object B of $\mathcal{K}$, and for all $f \in \mathcal{K}(A,B)$ and $g \in \mathcal{K}(B,A)$ we have

$$A \xrightarrow{id_A} A \xrightarrow{f} B = A \xrightarrow{f} B \quad \text{and} \quad B \xrightarrow{g} A \xrightarrow{id_A} A = B \xrightarrow{g} A.$$

Examples:

Set = <Sets and Maps>

Pfn = <Sets and Partial functions>

Rel = <Sets and Relations>

Vect = <Vector spaces & linear maps>

Grp = <Groups and Homomorphisms>

Mon = <Monoids and Homomorphisms>

Met = <Metric spaces and continuous maps>

and Top = <Topological spaces and continuous maps>

Given a poset $(P,\leq)$, we may associate with it a category which has one object p for each element p of P, and such that there is a (unique) morphism from p to p' iff $p \leq p'$. [Since $p \leq p$, there is an identity morphism for each p, etc.]

1.2 Given a category $\mathcal{K}$ we define its opposite (or dual) category $\mathcal{K}^{op}$ to have $Obj(\mathcal{K}^{op}) = Obj(\mathcal{K})$ while

$$\mathcal{K}^{op}(A,B) = \{B \xleftarrow{f} A \mid f \in \mathcal{K}(A,B)\}$$

with composition defined by $A \xleftarrow{f} B \xleftarrow{g} C = C \xrightarrow{f \cdot g} A$ and identities $A \xleftarrow{id_A} A$ as in $\mathcal{K}$. [Since the axioms of I and II of 1.1 are preserved under arrow-reversal it is immediate that $\mathcal{K}^{op}$ is indeed a category.]

1.3 DUALITY PRINCIPLE FOR CATEGORY THEORY: Let W be any construct defined for any category $\mathcal{K}$. Then the dual of W, called co-W, is the construct defined for any category $\mathcal{K}$ by defining W in $\mathcal{K}^{op}$ (and, hence, reversing all the arrows, from the point of view of $\mathcal{K}$).

If T is a theorem true for all categories $\mathcal{K}$, then the dual of T, obtained by reversing all the arrows of T, is true for all categories $\mathcal{K}^{op}$, and so (since $(\mathcal{K}^{op})^{op} = \mathcal{K}$) is true for all categories.

In other words, duality "cuts the work in half".

1.4 DEFINITION: A *semiadditive category* is a category $\mathcal{K}$ together with an abelian semigroup structure on each of its morphism sets, subject to the conditions

(a) Composition is bilinear; i.e., whenever $f\cdot g$ and $f\cdot h$ are defined we have

$$f\cdot g + f\cdot h = f\cdot(g+h)$$

and whenever $g\cdot f$ and $h\cdot f$ are defined we have

$$g\cdot f + h\cdot f = (g+h)\cdot f$$

(b) $(g : A \to B)\cdot(0 : B \to C) = 0 : A \to C$; similarly, $0\cdot g = 0$.

If, moreover, each $\mathcal{K}(A,B)$ is an abelian group we call $\mathcal{K}$ an *additive category*.

Just as a one-object category is a monoid, so is a one-object semiadditive category a *semiring*, and a one-object additive category is a *ring*.

2. PRODUCTS AND LIMITS

2.1 A *product* in the category $\mathcal{K}$ of a family of objects $\{A_i \mid i \in I\}$ is an object A equipped with a family of morphisms $\{\pi_i : A \to A_i \mid i \in I\}$ (called *projections*) with the universal property that, given any other object C equipped with a family of morphisms $\{p_i : C \to A_i \mid i \in I\}$ there exists a unique morphism p such that

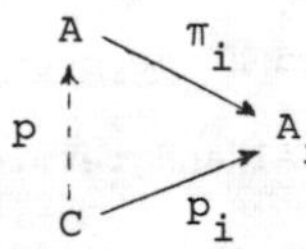

for all $i \in I$. (2)

We write $p = (p_i)$ and $A = \prod_{i\in I} A_i$. If $I = \{1,2,\ldots,n\}$ we may write $A = A_1 \times A_2 \times \cdots \times A_n$.

In <u>Set</u>, this is just the cartesian product. Again, the disjoint union of <u>Set</u> generalizes to:

<u>2.2</u> The family of morphisms $\{in_i : A_i \to A \mid i \in I\}$ (called <u>injections</u>) is called a <u>coproduct</u> in $\mathcal{K}$ iff it is a product in $\mathcal{K}^{op}$. If the family q_i induces the morphism q, we write $q = (q_i)$; and we write $A = \coprod_{i \in A} A_i$. If $I = \{1,2,\ldots,n\}$ we may write $A = A_1+A_2+\ldots+A_n$.

The product of vector spaces is a cartesian product, just as in <u>Set</u>, but the coproduct is strikingly different from the disjoint union, being the weak direct sum.

In the poset-cum-category $(P,\leq)$, the diagram (2) says

$$a \leq a_i \text{ for all } i, \text{ while } c \leq a_i \text{ for all } i \Rightarrow c \leq a.$$

In other words, a product in $(P,\leq)$ is a <u>greatest lower bound</u>; and so arbitrary collections of objects in (elements of) P have a product iff every subset of P has a greatest lower bound.

Dually, a coproduct is a <u>least upper bound</u>.

For example, every collection of objects in $(\mathcal{P}(S),\subset)$ has both a product and a coproduct:

$$\coprod_{i \in I} A_i = \bigcup_{i \in I} A_i \quad \text{and} \quad \prod_{i \in I} A_i = \bigcap_{i \in I} A_i$$

so that <u>unions</u> and <u>intersections</u> have a pleasing categorical interpretation.

<u>2.3</u> <u>DEFINITION</u>: An object T in the category $\mathcal{K}$ is <u>terminal</u> iff for every object A of $\mathcal{K}$ there is a unique morphism $A \to T$. Dually, an object I in the category $\mathcal{K}$ is <u>initial</u> iff for every object A of $\mathcal{K}$ there is a unique morphism $I \to A$.

An object 0 of a category $\mathcal{K}$ is called a <u>zero object</u> (compare the object $\{0\}$ of <u>Vect</u>) if it is both initial and terminal; i.e., for objects A and B, there is exactly one morphism $A \to 0$ and $0 \to B$.

2.4 DEFINITION: Given $f_1 : A_1 \to A$ and $f_2 : A_2 \to A$, we say the commutative diagram

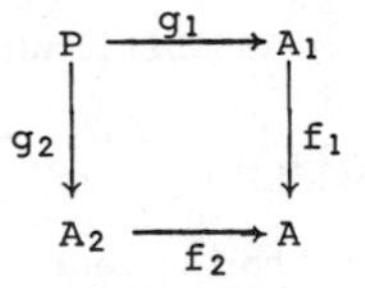

is a pullback of f_1 and f_2 if it has the property that any commutative diagram

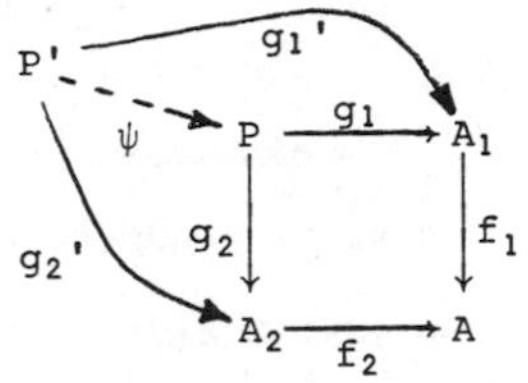

can be completed by a unique ψ.

The dual construction is called a pushout.

2.5 DEFINITION: Let $\mathcal{K}$ be a category with zero object 0. Then the kernel $u : K \to A$ of $f : A \to B$ is given by the pullback diagram

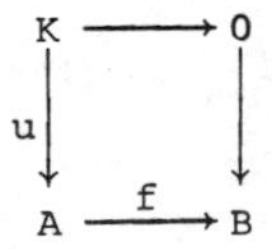

To provide a general framework for these concepts, we first gain some motivation from lattice theory:

Consider a partially ordered set such that every subset D of L has a least upper bound $\sup D \in L$.

Set $\perp = \sup \emptyset$, where $\emptyset$ is the empty subset of L.

By the definition of sup, it follows that $\perp \leq a$ for every $a \in L$ (since every a is an upper bound for $\emptyset$, and $\perp$ is the least such).

Given any subset D, form the set $\hat{D} = \{\hat{d} \mid \hat{d} \leq a \text{ for every } a \text{ in } D\}$

Let us show that the existence of lub's allows us to define glb's by:

$$\text{glb } D = \sup \hat{D}$$

$$a = \sup \hat{D} \iff \hat{d} \leq a \text{ for every } \hat{d} \in \hat{D} \text{ and if } \hat{d} \leq a' \text{ for all } \hat{d} \in \hat{D} \text{ then } a \leq a'$$

$$\Longrightarrow a \leq a' \text{ for every } a' \in D.$$

Thus a is a lower bound for D. Suppose $\bar{a}$ were another lower bound. Then $\bar{a}$ would be in $\hat{D}$, and so we have $\bar{a} \leq a$, which is thus the greatest lower bound. Hence the following definitions:

2.6 DEFINITION: A complete lattice is a partially ordered set $(L,\leq)$ such that every subset D of L has a least upper bound, $\sup D$. We set $\perp = \sup \phi$; $\top = \sup L$.

We say D is directed if it contains the $\sup$ of all its finite subsets.

We say $f : (L,\leq) \to (L',\leq)$ is continuous if, whenever $D \subset L$ is directed,

$$f(\sup D) = \sup f(D).$$

[Thus f is monotonic: $x \leq y \Rightarrow f(x) \leq f(y)$.]

2.7 PROPERTY: Complete lattices, together with continuous functions, form a category, CLatt. [Key Lemma: D directed, f continuous $\Rightarrow f(D)$ directed.]

We now use these ideas to motivate some general categorical definitions: Recall that a poset $(P,\leq)$ may be viewed as a category with an object for each element p of P, and with a morphism $p \to p'$ just in case $p \leq p'$. Thus x is a lower bound for $S \subset P$ iff there exists a morphism $x \to s$ for each s in S; and x is a glb for S if it is terminal in the subcategory of P comprising the lower bounds of S. Generalizing these ideas to arbitrary categories, we have:

2.8 DEFINITION: A diagram D in a category $\mathcal{K}$ is a directed graph whose vertices i are labelled by objects D_i of $\mathcal{K}$ and whose edges $i \to j$ (several may link a given i and j) are labelled by morphisms in $\mathcal{K}(D_i,D_j)$.

A cone for a diagram D is a family $X \to D_i$ of morphisms from a single object X such that $X \to D_i \to D_j = X \to D_j$ for every $D_i \to D_j$ in D. A morphism from a cone $(X \to D_i)$ to a cone $(X' \to D_i)$ is a $\mathcal{K}$-morphism $X \to X'$ such that $X \to X' \to D_i = X \to D_i$ for all i. The cones for D then form a category, and a limit for the diagram D is a terminal object in this category, i.e., $(X \to D_i)$ is such that for all cones $(X' \to D_i)$ on D there is a unique morphism $X \to X'$ such that $X' \to X \to D_i = X' \to D_i$.

We say a family $(D_i \to X)$ is a colimit for D if it is a limit for D considered in $\mathcal{K}^{op}$.

2.9 OBSERVATION: Since terminal objects are unique up to isomorphism, so too are limits and colimits.

Note that in the poset category $(P,\leq)$ a diagram D is equivalent to a subset of P, and that the limit of D is then the glb, while the colimit is a lub.

2.10 EXAMPLES:

1. $$\begin{array}{ccc} & A \times B & \\ {}^{p_1}\swarrow & & \searrow{}^{p_2} \\ A & & B \end{array}$$ is the limit of the diagram $A \quad B$ (no arrows); while the coproduct is the colimit of the diagram.

2. The limit of $$\begin{array}{ccc} & & A_2 \\ & & \downarrow \\ A_1 & \longrightarrow & A \end{array}$$ is its pullback; while the colimit of $$\begin{array}{ccc} A & \longrightarrow & A_1 \\ \downarrow & & \\ A_2 & & \end{array}$$ is its pushout.

3. (See also section 3) $X \longrightarrow A \underset{f_2}{\overset{f_1}{\rightrightarrows}} B$, the equalizer, is the limit of $A \underset{f_2}{\overset{f_1}{\rightrightarrows}} B$; while its colimit is the coequalizer.

2.11 OBSERVATION: Given a diagram D in Set with typical edge $D_i \xrightarrow{f} D_j$, we may form the limit of D by setting (a typical equalizer construction)

$$X = \{\text{sequences} \quad (x_i) \in \Pi D_i \mid f(x_i) = x_j \quad \text{for all} \quad f \in D\}$$

and then define

$$X \to D_i : (x_i) \mapsto x_i .$$

Clearly, each $X \to D_i \xrightarrow{f} D_j = X \to D_j$, so that $(X \to D_i)$ is a cone; and it is routine to check that it is terminal.

This corresponds to the more general result:

2.12 THEOREM: If a category $\mathcal{K}$ has a terminal object, equalizers of all pairs of morphisms, and products of all pairs of objects, then $\mathcal{K}$ is finite complete (i.e., has limits of all diagrams indexed by finite sets).

If $\mathcal{K}$ has equalizers of all pairs of arrows and all small products (i.e., products of sets of objects), then $\mathcal{K}$ is small-complete (i.e., contains limits of all diagrams indexed by sets).

<u>2.13</u> <u>DEFINITION</u>: A category $\mathcal{K}$ is <u>filtered</u> when $\mathcal{K}$ is nonempty and

(i) For any two objects A_1 and A_2 there exists a diagram

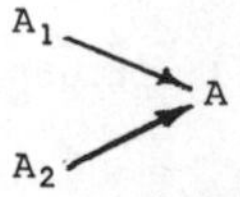

(ii) Given any two morphisms $f_1, f_2 : A \to B$ with common domain and codomain, there exists a morphism $g : B \to C$ such that

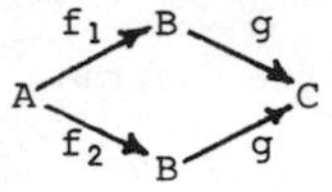

commutes.

Thus any finite diagram in a filtered category $\mathcal{K}$ is the base of at least one cone with a vertex $K \in \mathcal{K}$.

<u>2.14</u> <u>DEFINITION</u>: A <u>filtered colimit</u> is a limit of a functor $F : \mathcal{K} \to \mathcal{L}$ (qua diagram) defined on a filtered category $\mathcal{K}$.

3. EPIMORPHISMS AND MONOMORPHISMS

<u>3.1</u> <u>DEFINITION</u>: An <u>epimorphism</u> in a category $\mathcal{K}$ is a $\mathcal{K}$-morphism $f : A \to B$ with the property that two morphisms $g,h : B \to C$ can satisfy $g \cdot f = h \cdot f$ iff $g = h$.

$$A \xrightarrow{f} B \underset{h}{\overset{g}{\rightrightarrows}} C.$$

A <u>monomorphism</u> in a category $\mathcal{K}$ is a $\mathcal{K}$-morphism $f : B \to A$ with the property that two morphisms $g,h : C \to B$ can satisfy $f \cdot g = f \cdot h$ iff $g = h$.

$$C \underset{h}{\overset{g}{\rightrightarrows}} B \xrightarrow{f} A$$

Let us observe immediately that, in <u>Set</u> at least, these reduce to onto and one-to-one maps respectively.

<u>3.2</u> <u>EXAMPLE</u>: $f : A \to B$ is onto in <u>Set</u> if and only if every b is $f(a)$ for some a in A. Thus, if $gf(a) = hf(a)$ for every a in A, then $g(b) = h(b)$ for every b in B, and so $g = h$. Thus f is an epimorphism.

If $f : A \to B$ is not onto in Set, there is a b, call it b_1, in B which is not an $f(a)$. Set $g = id_B$, and define h so that $h(b) = b$ for $b \neq b_1$, while $h(b_1) \neq b_1$. Then $gf = hf$ even though $g \neq h$. Thus f is not an epimorphism.

$f : B \to A$ is one-to-one in Set $\Longleftrightarrow b_1 \neq b_2$ implies $f(b_1) \neq f(b_2)$ for all b_1, b_2 in B. Thus if $fg(c) = fh(c)$ we must have $g(c) = h(c)$ and so f is a monomorphism.

If $f : B \to A$ is not one-to-one, there must be $b_1 \neq b_2$ in B with $f(b_1) = f(b_2)$. Set $g = id_B$ and define h so that $h(b) = b$ for $b \neq b_1$, while $h(b_1) = b_2$. Then $fg = fh$ even though $g \neq h$. Thus f is not a monomorphism.

Note that the definitions of epimorphism and monomorphism are dual in the category-theoretic sense that one formal definition is obtained from the other by reversing the arrows--a duality that is obscured in the usual definitions of one-to-one and onto.

It is easy to show that in the categories R-Mod and Gp monomorphisms coincide with one-to-one homomorphisms. It can also be shown--though the proof for Gp is not obvious--that the epimorphisms in these categories are precisely the onto homomorphisms. Despite these preliminary successes,* we adopt the philosophy that more general categories than these will be useful in system theory; and it has been gradually discovered by category theorists that no one concept of "epimorphism" or "monomorphism" is adequate--rather, it is profitable to axiomatize a class of possibilities.

---

*In the category of monoids and monoid homomorphisms the inclusion map of the natural numbers into the integers is an epimorphism which is not onto.

<u>3.3</u> <u>DEFINITION</u>: An <u>image factorization system</u> for a category $\mathcal{K}$ consists of a pair $(\mathcal{E},\mathcal{M})$ where $\mathcal{E}$ and $\mathcal{M}$ are classes of morphisms in $\mathcal{K}$ satisfying the following four axioms:

<u>IFS 1</u>: If $e : A \to B \in \mathcal{E}$ and $e' : B \to C \in \mathcal{E}$ then $e'e : A \to C \in \mathcal{E}$. Dually, if $m : A \to B \in \mathcal{M}$ and $m' : B \to C \in \mathcal{M}$ then $m'm : A \to C \in \mathcal{M}$.

<u>IFS 2</u>: If $e : A \to B \in \mathcal{E}$, e is an epimorphism. Dually, if $m : A \to B \in \mathcal{M}$, m is a monomorphism.

<u>IFS 3</u>: If $f : A \to B$ is an isomorphism then $f \in \mathcal{E}$ and $f \in \mathcal{M}$.

<u>IFS 4</u>: Every $f : A \to B$ in $\mathcal{K}$ has an $\mathcal{E}$-$\mathcal{M}$ factorization which is unique up to isomorphism. More precisely, there exists an <u>$\mathcal{E}$-$\mathcal{M}$ factorization</u> (e,m) of f, meaning $e \in \mathcal{E}$, $m \in \mathcal{M}$ and $f = me$, (so that there exists an object--call it $f(A)$--such that e has the form $e : A \to f(A)$ and m has the form $m : f(A) \to B)$, and this factorization is unique in the sense that if (e',m')

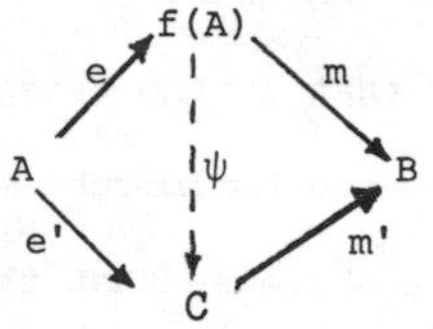

is another such factorization -- $f = m'e'$, $e' \in \mathcal{E}$, $m' \in \mathcal{M}$ -- then there exists an isomorphism ψ (as shown above) with $\psi e = e'$, $m'\psi = m$.

In the category <u>Set</u> of sets,

$$\mathcal{E} = \{\text{onto functions}\} \quad \text{and} \quad \mathcal{M} = \{\text{one-to-one functions}\}$$

is an image factorization system. The first three axioms are clear. For IFS 4, define $f(A) = \{f(a) : a \in A\} \subset B$, set m to be the inclusion function and define $e(a) = f(a) \in I$. For the uniqueness proof, define $\psi(i) = e'(a)$ for any a with $e(a) = i$. The remaining details are routine. Essentially the same construction demonstrates that

$$\mathcal{E} = \{\text{onto homomorphisms}\} \quad \text{and} \quad \mathcal{M} = \{\text{one-to-one homomorphisms}\}$$

provides image factorizations in <u>R-Mod</u> and <u>Grp</u>.

3.4 DIAGONAL FILL-IN LEMMA: Given a commutative square -- $ge = mf$ -- as shown below with

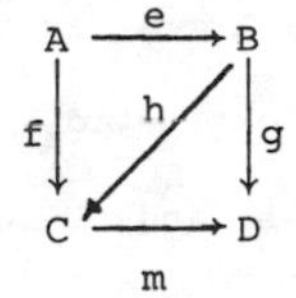

$e \in \mathcal{E}$ and $m \in \mathcal{M}$ there exists a unique h (as shown) with $he = f$, $mh = g$. The proof uses the following diagram:

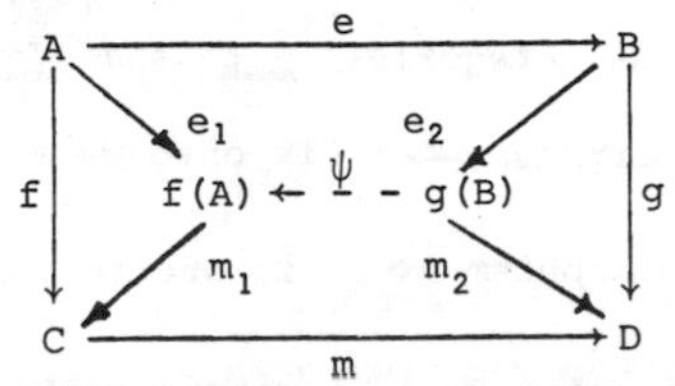

3.5 DEFINITION: A morphism $A \xrightarrow{h} B$ is a coequalizer iff there exists a pair $p_1, p_2 : R \to A$ of morphisms such that $h \cdot p_1 = h \cdot p_2$, and such that whenever $A \xrightarrow{h'} B'$ satisfies $h' \cdot p_1 = h' \cdot p_2$, there is a unique morphism† $B \xrightarrow{\psi} B'$ such that $\psi \cdot h = h'$.

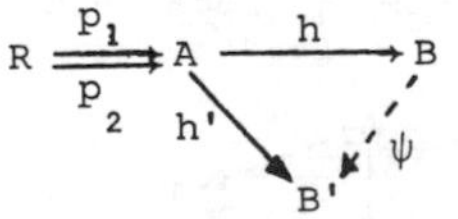

In this situation, we call h the coequalizer of p_1 and p_2 and write $h = \mathrm{coeq}(p_1, p_2)$.

3.6 DEFINITION: A morphism $B \xrightarrow{h} A$ is an equalizer in $\mathcal{K}$ if it is a coequalizer in $\mathcal{K}^{op}$. If $h = \mathrm{coeq}(q_1, q_2)$ in $\mathcal{K}^{op}$, we write $h = \mathrm{eq}(q_1, q_2)$ in $\mathcal{K}$, and say h is the equalizer of q_1 and q_2.

3.7 PROPOSITION: Every coequalizer is an epimorphism.

Proof: If $h = \mathrm{coeq}(p_1, p_2)$ and $k_1 \cdot h = k_2 \cdot h$, then

$$(k_1 \cdot h) \cdot p_1 = k_1 \cdot (h \cdot p_1) = k_1 \cdot (h \cdot p_2) = (k_1 \cdot h) \cdot p_2 \qquad (1)$$

and so there is a unique morphism ϕ such that $k_1 \cdot h = \phi \cdot h$. Thus $k_1 = k_2$. □

---

†Unique in the category $\mathcal{K}$ with which we are working, of course.

<u>3.8</u> <u>PROPOSITION</u>: Every equalizer is a monomorphism. □

<u>3.9</u> <u>DEFINITION</u>: A morphism $f : A \to B$ is an <u>isomorphism</u> iff there exists a morphism $k : B \to A$ such that $k \cdot f = id_A$ and $f \cdot k = id_B$. We call such a k an <u>inverse</u> of f. We say A and B are <u>isomorphic</u>, $A \cong B$, just in case $\mathcal{K}(A,B)$ contains an isomorphism.

<u>3.10</u> <u>FACT</u>: An isomorphism $f : A \to B$ is both an equalizer and a coequalizer.

<u>3.11</u> <u>PROPOSITION</u>: in the categories <u>Set</u> and <u>Vect</u>, we have

(i) f is an epimorphism $\Longleftrightarrow$ f is onto $\Longleftrightarrow$ f is a coequalizer

(ii) f is a monomorphism $\Longleftrightarrow$ f is one-to-one $\Longleftrightarrow$ f is an equalizer.

<u>3.12</u> <u>DEFINITION</u>: If $f : A \to B$ is a monomorphism we call A a <u>subobject</u>* of B, and refer to f as the <u>inclusion</u> of A in B. Sometimes we omit mention of f, and write $A \subset B$, and say that B <u>contains</u> A. Dually, if $f : A \to B$ is an epimorphism, we call B a <u>quotient object</u> of A.

If $f_1 : A_1 \to A$ and $f_2 : A_2 \to A$ are monomorphisms, we write $f_1 \leq f_2$ just in case there exists γ such that $f_2\gamma = f_1$:

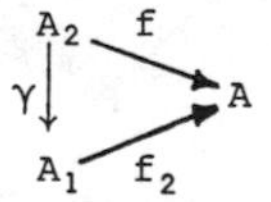

Clearly γ must be a monomorphism, and if $f_1 \leq f_2$ and $f_2 \leq f_1$ then A_1 and A_2 are isomorphic. Dually, we write $f_1 \leq f_2$ for epimorphisms iff $f_1 \leq f_2$ as monomorphisms in $\mathcal{K}^{op}$.

<u>3.13</u> <u>DEFINITION</u>: Let $\{f_i : A_i \to A \mid i \in I\}$ be a family of subobjects of A. Then the <u>intersection</u> of the family is a greatest lower bound in the $\leq$ ordering. i.e., $f : A' \to B$ satisfies $f \leq f_i$ for all i; and if $f' \leq f_i$ for all i, then $f' \leq f$. We write $\bigcap_{i \in I} A_i$ for A'.

---

*It is equally common for a subobject of B to mean an antisymmetry equivalence class of B-valued monomorphisms.

3.14 PROPOSITION: If $A_1 \to A$ and $A_2 \to A$ are monomorphisms in a category $\mathcal{K}$, then the diagram

$$\begin{array}{ccc} P & \longrightarrow & A_1 \\ \downarrow & & \downarrow \\ A_2 & \longrightarrow & A \end{array}$$

is a pullback iff $P \to A_1 \to A = P \to A_2 \to A$ is the intersection of A_1 and A_2. Hence [since $\bigcap_{i=1}^{n} A_i = \left(\bigcap_{i=1}^{n-1} A_i\right) \cap A_n$] if $\mathcal{K}$ has pullbacks then $\mathcal{K}$ has finite intersections.

3.15 DEFINITION: $f : A \to B$ is called a *retraction* if there exists a morphism $g : B \to A$ such that $fg = 1_B$; and we then say that g is a *coretraction*, and that B is a *retract* of A.

Note that f is an isomorphism iff it is both a retraction and a coretraction. It is clear that if f is a retraction it is an epimorphism ($h_1 f = h_2 f$ implies $h_1 = h_1 1_B = h_1 fg = h_2 fg = h_2$)--retractions are thus also called *split epimorphisms*. Coretractions (which are monomorphisms) are also called *sections*, as motivated by the figure for the case of, e.g., *Set*:

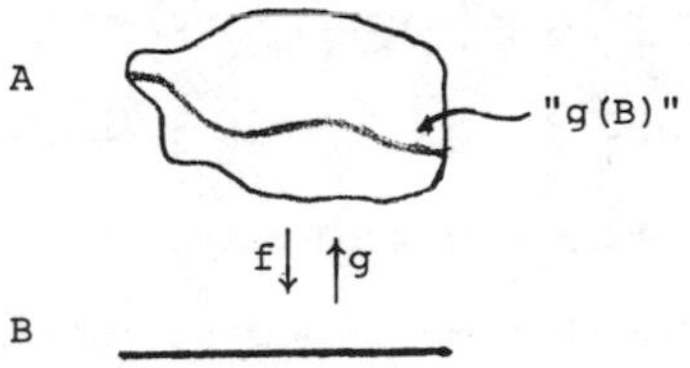

4. FUNCTORS AND ADJOINTS

4.1 DEFINITION: A *functor* H from a category $\mathcal{K}$ to a category $\mathcal{L}$ is a function which maps $\mathrm{Obj}(\mathcal{K}) \to \mathrm{Obj}(\mathcal{L}) : A \mapsto HA$, and which for each pair A,B of objects of $\mathcal{K}$ maps $\mathcal{K}(A,B) \to \mathcal{L}(HA,HB) : f \mapsto Hf$, while satisfying the two conditions:

$$H\, id_A = id_{HA} \quad \text{for every } A \in \mathrm{Obj}(\mathcal{K})$$

$$H(f \cdot g) = Hf \cdot Hg \quad \text{whenever } f \cdot g \text{ is defined in } \mathcal{K}.$$

We say H is an *isomorphism* if $A \mapsto HA$ and $\mathcal{K}(A,B) \to \mathcal{L}(HA,HB)$ are bijections.

The most trivial example of a functor (or of an isomorphism), is, for each category $\mathcal{K}$, the identity functor $id_{\mathcal{K}} : \mathcal{K} \to \mathcal{K}$ which sends A to A and f to f.

We also observe that, for any set X_0, $-\times X_0 : \underline{Set} \to \underline{Set}$ is a functor:

$$id_Q \times X_0 = id_{Q\times X_0} : Q\times X_0 \to Q\times X_0 : (q,x) \mapsto (q,x)$$

while for $f : Q \to Q'$ and $g : Q' \to Q''$ we have

$$(g\cdot f)\times X_0 = (g\times X_0)\cdot(f\times X_0) : Q\times X_0 \to Q''\times X_0 : (q,x) \mapsto (g(f(q)),x).$$

A general and important class of functors are the forgetful functors. For example, monoids are sets with extra structure and homomorphisms are maps which satisfy extra conditions. Thus we have a well-defined mapping

$$U : \underline{Mon} \to \underline{Set}$$

which sends a monoid to its underlying set, and forgets that a homomorphism has anything special about it. Thus, $U\, id_M = id_M$ and $U(f\cdot g) = f\cdot g$ and so U is certainly a functor. Similarly, we can define forgetful functors $\underline{Gp} \to \underline{Mon}$, $\underline{Gp} \to \underline{Set}$, $\underline{Vect} \to \underline{Gp}$, $\underline{Met} \to \underline{Top}$, etc., etc., etc.

4.2 DEFINITION: Given two categories $\mathcal{K}$ and $\mathcal{L}$, we define their product $\mathcal{K}\times\mathcal{L}$ to be the category whose objects are ordered pairs (K,L) of objects K from $\mathcal{K}$ and L from $\mathcal{L}$, and for which morphisms in $(\mathcal{K}\times\mathcal{L})[(K,L),(K',L')]$ are just pairs (f,g) with $f \in \mathcal{K}(K,K')$ and $g \in \mathcal{L}(L,L')$, while $id_{(K,L)} = (id_K, id_L)$ and $(f',g')\cdot(f,g) = (f'\cdot f, g'\cdot g)$.

4.3 OBSERVATION: The map which assigns to each pair of objects K,K' in the category $\mathcal{K}$ the set $\mathcal{K}(K,K')$ of morphisms from K to K' becomes a functor

$$hom : \mathcal{K}^{op}\times\mathcal{K} \to \underline{Set} : (K,K') \mapsto \mathcal{K}(K,K')$$

when we make the morphism assignment

$$(K_1 \xleftarrow{f} K_1', K_2 \xrightarrow{g} K_2'') \mapsto \mathcal{K}(K_1,K_2) \xrightarrow{g\cdot(\ )\cdot f} \mathcal{K}(K_1',K_2').$$

[hom is also known as the external representation functor.]

4.4 The *free monoid on the set* X *of generators* is the set X^* of all finite sequences of elements from X (including the "empty" sequence Λ of length 0) with the associative multiplication of *concatenation*

$$(x_1,\ldots,x_m)\cdot(x_1',\ldots,x_n') = (x_1,\ldots,x_m, x_1',\ldots,x_n')$$

for which Λ is clearly the identity: $\Lambda\cdot w = w = w\cdot\Lambda$ for all $w \in X^*$.

Quite apart from its fundamental role in automata theory and formal language theory, $(X^*, \mathrm{conc}, \Lambda)$ is interesting because of the following property:

Given any monoid $(S,\circ,e)$ and any map f from X to S, there is a unique homomorphism ψ from $(X^*, \mathrm{conc}, \Lambda)$ to $(S,\circ,e)$ which extends f, i.e., such that $\psi((x)) = f(x)$ for each x in X; $\psi(w\cdot w') = \psi(w)\circ\psi(w')$ for all w,w' in X^*; and $\psi(\Lambda) = e$. In fact, it is clear that the one and only ψ is defined by

$$\psi(\Lambda) = e$$

$$\psi((x_1,\ldots,x_n)) = f(x_1)\circ\ldots\circ f(x_n)$$

Using $\eta : X \to X^*$ to denote the "inclusion of generators" map $x \to (x)$, we may express the situation in the following categorical form:

4.5 The monoid $A = (X^*, \mathrm{conc}, \Lambda)$ [for which $UA = X^*$, where $U : \mathbf{Mon} \to \mathbf{Set}$ is the forgetful functor] is equipped with a map $\eta : X \to UA$ in such a way that, given any other monoid M and any map $f : X \to UM$, there exists a unique *homomorphism* ψ such that

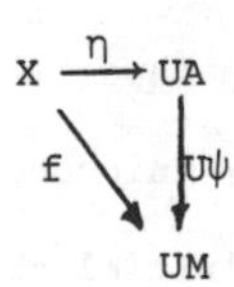

The argument for uniqueness up to isomorphism of limits and colimits can be adapted to yield:

4.6 LEMMA: A monoid A equipped with a map $\eta : X \to UA$ satisfies the condition 4.5 iff there exists a monoid isomorphism $\psi : A \to (X^*, \mathrm{conc}, \Lambda)$ such that $(U\psi\cdot\eta)(x) = (x)$ for all x in X. □

Thus we have succeeded in taking an element-by-element definition and showing that it can be captured up to isomorphism in arrow-theoretic terms. This leads to the following general definition:

4.7 DEFINITION: Let $G : \mathcal{A} \to \mathcal{B}$ be any functor, and B an object of $\mathcal{B}$. We say the pair (A, η), where A is an object of $\mathcal{A}$ and $\eta : B \to GA$ is a morphism of $\mathcal{B}$, is <u>free over</u> B <u>with respect to</u> G just in case $\eta : B \to GA$ has the couniversal property that given any morphism $f : B \to GA'$ with A' an object of $\mathcal{A}$, there exists a unique $\mathcal{A}$-morphism $\psi : A \to A'$ such that

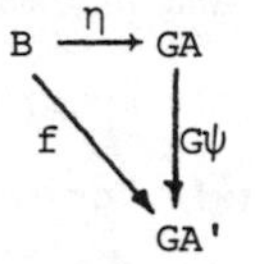

$$\tag{1}$$

We refer to η as the <u>inclusion of generators</u>; and call the unique ψ satisfying (1) the $\mathcal{A}$-<u>morphic extension of</u> ψ (with respect to G). As in <u>4.6</u>, if (A, η) and (A', η') are both free over B w.r.t. G, then $G\psi \cdot \eta = \eta'$ for some isomorphism $\psi : A \to A'$. However, it may be instructive to give a quite different proof: Let $(B \downarrow G)$ be the category whose objects are $\mathcal{B}$-morphisms $B \xrightarrow{\eta} GA$ for $\mathcal{A}$-objects A, while the morphisms are $\mathcal{A}$-morphisms $\psi : A \to A'$ such that

$$
\begin{array}{ccc}
 & B & \\
\eta \swarrow & & \searrow \eta' \\
GA & \xrightarrow[G\psi]{} & GA'
\end{array}
$$

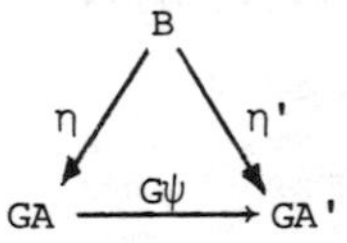

commutes. We immediately recognize that (A, η) is <u>free</u> over B iff $B \xrightarrow{\eta} GA$ is <u>initial</u> in $(B \downarrow G)$ -- uniqueness up to isomorphism of free objects then follows from that of initial objects.

4.8 DEFINITION: Let $F : \mathcal{A} \to \mathcal{B}$ be any functor, and B an object of $\mathcal{B}$. We say the pair (A, ε) where A is an object of $\mathcal{A}$ and $\varepsilon : FA \to B$ is a morphism of $\mathcal{B}$, is <u>cofree</u> over B <u>with respect to</u> F just in case $\varepsilon : FA \to B$ has the universal property that given any morphism $f : FA' \to B$

with A' an object of $\mathcal{A}$, there exists a unique $\mathcal{A}$-morphism $\psi : A' \to A$ such that

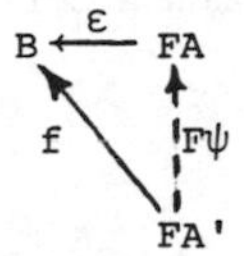

For example, let F be the functor $-\times X_0 : \underline{\underline{Set}} \to \underline{\underline{Set}}$. Then for each B we may take the corresponding (A,ε) to be given by

$$A = B^{X_0}, \text{ the set of all maps from } X_0 \text{ to } B$$

$$\varepsilon : B^{X_0} \times X_0 \to B : (f,x) \mapsto f(x) \text{ is evaluation.}$$

Then given $f : A' \times X_0 \to B$, the corresponding ψ is clearly

$$\psi : A' \to B^{X_0} : a \mapsto f(a,\cdot)$$

where $f(a,\cdot) : X_0 \to B : x \mapsto f(a,x)$.

If every B has a free (A,η) with respect to the functor $G : \mathcal{A} \to \mathcal{B}$, then we can introduce a functor $F : \mathcal{B} \to \mathcal{A}$ for which FB is free over B. Such a functor F is called a *left adjoint* of G. Dually, if every B has a cofree object, then there is a cofree-object choice functor called the *right adjoint* of G. Thus we say that G *has a left adjoint* just in case every object B has a pair (A,η) free over B; and that F *has a right adjoint* if every A has a pair (B,ε) cofree over A.

4.9 THEOREM: Let $G : \mathcal{A} \to \mathcal{B}$ be a functor with the property that to every B in $\mathcal{B}$ there corresponds a free object, call it $(FB,\eta B)$ [so that $\eta B : B \to GFB$]. Given any $f : B \to B'$ in $\mathcal{B}$, define a morphism $Ff : FB \to FB'$ by the diagram

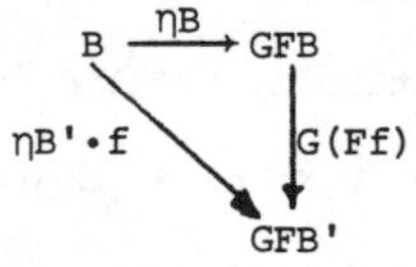

Then the collection of maps $F : \mathcal{B} \to \mathcal{A}$ so defined is a functor (clearly the object map is unique up to isomorphism, and fixes the morphism maps uniquely) called the *left adjoint* of G.

Dually, we have:

4.10 THEOREM: Let $F : \mathcal{B} \to \mathcal{A}$ be a functor with the property that to every A in $\mathcal{A}$ there corresponds a cofree object, call it $(GA, \varepsilon A)$ [so that $\varepsilon A : FGA \to A]$. Given any $f : A \to A'$ in $\mathcal{A}$, define a morphism Gf : GA \to GA' by the diagram

$$A \xleftarrow{\varepsilon A} FGA$$
$$f \cdot \varepsilon A' \qquad \qquad \Big\uparrow F(gf)$$
$$FGA'$$

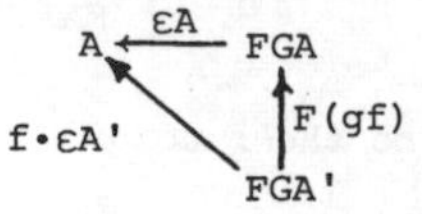

Then $G : \mathcal{A} \to \mathcal{B}$ is a functor, which we call the right adjoint of F. □

4.11 DEFINITION: A subcategory $\mathcal{L}$ of $\mathcal{K}$ is called reflective in $\mathcal{K}$ when the inclusion functor $F : \mathcal{L} \to \mathcal{K}$ has a left adjoint $G : \mathcal{K} \to \mathcal{L}$. Thus for every $f : K \to FL$ (with $K \in \mathcal{K}$, $L \in \mathcal{L}$) there exists unique $\psi : GK \to L \in \mathcal{L}$ with $F\psi \cdot \eta K = f$.

$$K \xrightarrow{\eta K} FGK$$
$$f \searrow \qquad \Big\downarrow F\psi$$
$$FL$$

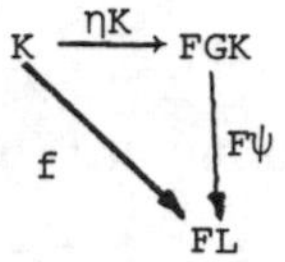

4.12 DEFINITION: Given two functors F_1 and F_2 from $\mathcal{K}$ to $\mathcal{L}$, a natural transformation $\tau : F_1 \xrightarrow{\;\;\cdot\;\;} F_2$ is a function which assigns to each object K of $\mathcal{K}$ a morphism $\tau K : F_1 K \to F_2 K$ in such a way that every $\mathcal{K}$ -morphism $f : K_1 \to K_2$ of $\mathcal{K}$ yields a commutative diagram:

$$
\begin{array}{ccc}
K_1 & F_1 K_1 \xrightarrow{\;\tau K_1\;} F_2 K_1 \\
f \Big\downarrow & F_1 f \Big\downarrow \qquad \qquad \Big\downarrow F_2 f \\
K_2 & F_1 K_2 \xrightarrow{\;\tau K_2\;} F_2 K_2
\end{array}
$$

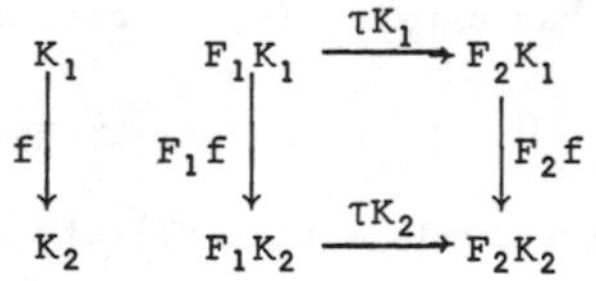

If we think of each functor F as giving a "representation" or "picture" of the category $\mathcal{K}$ in the category $\mathcal{L}$, then the definition says that the representations are naturally related in that we can find a single way of transforming the picture $F_1 K$ of an object into the picture $F_2 K$ of that same object which is consistent with every morphism which involves that object as domain or codomain.

4.13 DEFINITION: Let $\mathcal{A}$ and $\mathcal{B}$ be categories. An adjunction from $\mathcal{B}$ to $\mathcal{A}$ is a triple $<F, G, \phi>$ where F and G are functors

$$\mathcal{B} \underset{G}{\overset{F}{\rightleftarrows}} \mathcal{A}$$

and $\phi_{BA} : \mathcal{A}(FB,A) \cong \mathcal{B}(B,GA)$ is a <u>natural equivalence</u>, i.e., a natural transformation ϕ of functors $\mathcal{B}^{op} \times \mathcal{A} \to \underline{\underline{Set}}$ such that each ϕ_{BA} is bijective.

If $G : \mathcal{A} \to \mathcal{B}$ has F as a left adjoint, the universal property

$$\begin{array}{ccc} B & \xrightarrow{\eta} & GFB \\ & \searrow^{f} & \downarrow{\scriptstyle\psi} \\ & & GA \end{array}$$

establishes the adjunction $\langle F,G,\phi \rangle$ where $\phi_{BA}(\psi) = f$. Dually, an adjunction is obtained from a functor $F : \mathcal{B} \to \mathcal{A}$ with a right adjoint. Conversely, if (F,G,ϕ) is an adjunction G has F as a left adjoint and F has G as a right adjoint.

An equivalent way to define an adjunction from $\mathcal{B}$ to $\mathcal{A}$ is as a quadruple $\langle F,G,\eta,\varepsilon \rangle$ where F and G are functors

$$\mathcal{B} \underset{G}{\overset{F}{\leftrightarrows}} \mathcal{A}$$

and $\eta : id_{\mathcal{B}} \to GF$, $\varepsilon : FG \to id_{\mathcal{A}}$ are natural transformations, subject to the two <u>triangular identities</u>

$$\begin{array}{ccc} G & \xrightarrow{\eta G} & GFG \\ & \searrow^{id} & \downarrow{\scriptstyle G\varepsilon} \\ & & G \end{array} \qquad \begin{array}{ccc} F & \xrightarrow{\eta F} & FGF \\ & \searrow^{id} & \downarrow{\scriptstyle \varepsilon F} \\ & & F \end{array}$$

Given $\langle F,G,\phi \rangle$, $\eta_B : B \to GFB = \phi_{B,FB}(id_{FB})$ and $\varepsilon_A : FGA \to A = \phi^{-1}_{GA,A}(id_{GA})$.

Conversely, given $\langle F,G,\eta,\varepsilon \rangle$ define $\phi_{BA}(\psi : FB \to A) = B \xrightarrow{\eta B} GFB \xrightarrow{G\psi} GA$.

<u>LEMMA</u>: Given two adjunctions $\langle F,U,\eta,\varepsilon \rangle : \mathcal{A} \to \mathcal{B}$ and $\langle \bar{F},\bar{U},\bar{\eta},\bar{\varepsilon} \rangle : \mathcal{B} \to \mathcal{C}$, the composite functors yield an adjunction

$$\langle \bar{F}F, U\bar{U}, U\bar{\eta}F \cdot \eta, \bar{\varepsilon} \cdot \bar{F}\varepsilon\bar{U} \rangle : \mathcal{A} \to \mathcal{C}. \qquad \square$$

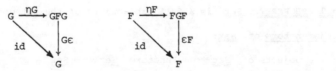

5. MONADS AS GENERALISED MONOIDS

A <u>monoid</u> is a set M together with an associative binary operator m and a nullary operator e which is an identity for M.

In other words, we have a diagram

$$M\times M \xrightarrow{m} M \xleftarrow{e} 1$$

satisfying the two conditions that the following diagrams commute:

<u>associativity</u>:

$$\begin{array}{ccc} (M\times M)\times M \cong M\times(M\times M) & \xrightarrow{id\times m} & M\times M \\ \downarrow{\scriptstyle m\times id} & & \downarrow{\scriptstyle m} \\ M\times M & \xrightarrow{\quad m \quad} & M \end{array}$$

<u>identity</u>:

$$\begin{array}{ccccc} M \cong 1\times M & \xrightarrow{e\times id} & M\times M & \xleftarrow{id\times e} & M\times 1 \cong M \\ & {\scriptstyle id}\searrow & \downarrow{\scriptstyle m} & \swarrow{\scriptstyle id} & \\ & & M & & \end{array}$$

Our task now is to consider constructs with the same defining conditions, but which live in categories other than **Set**. We note that we used the availability of a map × and an object 1 ε **Set** such that

Set × **Set** $\xrightarrow{\times}$ **Set** is associative up to isomorphism; and (1)

$1 \times \cdot \cong id \cong \cdot \times 1$ (2)

5.1 <u>PROPERTY</u>: **Set** is a full subcategory of the category $\mathbf{Set}^{\mathbf{Set}}$ of <u>endofunctors</u> of **Set**:

Objects of $\mathbf{Set}^{\mathbf{Set}}$: Functors $\mathbf{Set} \xrightarrow{H} \mathbf{Set}$

Maps of $\mathbf{Set}^{\mathbf{Set}}$: Natural transformations $H \xrightarrow{\alpha} H'$

i.e., α associates a map $HX \xrightarrow{\alpha X} HX'$ to each X ε **Set** in such a way that the following diagram commutes for every $X \xrightarrow{f} Y$ ε **Set**:

$$\begin{array}{ccc} HX & \xrightarrow{\alpha X} & HX' \\ \downarrow{\scriptstyle Hf} & & \downarrow{\scriptstyle Hf'} \\ HY & \xrightarrow{\alpha Y} & HY' \end{array}$$

with identities in composition defined in the obvious way.

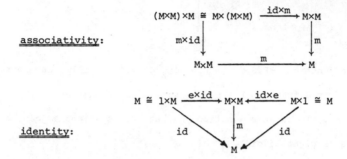

Proof: We embed $\mathbf{Set}$ in $\mathbf{Set}^{\mathbf{Set}}$ as a full subcategory, by sending X_0 to the endofunctor $\circ \times X_0$. We also send each $\mathbf{Set}$-map $f : X_0 \to Y_0$ to the $\mathbf{Set}^{\mathbf{Set}}$-map (i.e., natural transformation) $\circ\times X_0 \xrightarrow{id\times f} \circ\times Y_0$. That this is indeed a natural transformation follows from the commutativity of

$$\begin{array}{ccc} A\times X_0 & \xrightarrow{id_A\times f} & A\times Y_0 \\ \downarrow{\scriptstyle g\times id_{X_0}} & & \downarrow{\scriptstyle g\times id_{Y_0}} \\ B\times X_0 & \xrightarrow{id_B\times f} & B\times Y_0 \end{array}$$

Finally, to show that the image of $\mathbf{Set}$ is a full subcategory, we must show that any natural transformation $\circ\times X_0 \to \circ\times Y_0$ is of the form $id\times f$ for some $X_0 \xrightarrow{f} Y_0 \in \mathbf{Set}$. But we have the commutativity of

$$\begin{array}{ccc} A\times X_0 & \xrightarrow{\alpha A} & A\times Y_0 \\ \downarrow{\scriptstyle g\times id_{X_0}} & & \downarrow{\scriptstyle g\times id_{Y_0}} \\ B\times X_0 & \xrightarrow{\alpha B} & B\times Y_0 \end{array}$$

for all A, B and g. First pick $A = 1$ so that $1\times X_0 \xrightarrow{\alpha 1} 1\times Y_0$ may be regarded as a map $X_0 \xrightarrow{f} Y_0$. Then let $g(1) = b \in B$ to read from the diagram that

$$\begin{array}{ccc} (1,x) & \mapsto & (1,fy) \\ \downarrow & & \downarrow \\ (b,x) & \mapsto & (b,fy) \end{array}$$

so that $\alpha B = id_B \times f$. □

Thus, we may think of endofunctors in $\mathbf{Set}$ as generalizing sets. In fact, having gone this far, we may regard the endofunctors of any category $\mathcal{A}$ as our generalization of sets. Now we want to ensure that appropriate analogues of $\times$ and 1 **are available:**

We define $\mathcal{A}^{\mathcal{A}} \times \mathcal{A}^{\mathcal{A}} \to \mathcal{A}^{\mathcal{A}}$ to be simply **composition, given by**

$$(F,G) \to (\mathcal{A} \xrightarrow{F} \mathcal{A} \xrightarrow{G} \mathcal{A})$$

on objects. We must check that we can compose natural transformations appropriately:

<u>5.2 HORIZONTAL COMPOSITION:</u> $\left(\begin{array}{ccc} F & & G \\ \downarrow{\scriptstyle\alpha} & , & \downarrow{\scriptstyle\beta} \\ F' & & G' \end{array}\right) \longrightarrow \begin{array}{c} \mathcal{A} \xrightarrow{F} \mathcal{A} \xrightarrow{G} \mathcal{A} \\ \downarrow{\scriptstyle\beta\alpha} \\ \mathcal{A} \xrightarrow{F'} \mathcal{A} \xrightarrow{G'} \mathcal{A} \end{array}$

Two definitions present themselves:

$$GFS \xrightarrow{\beta FS} G'FS \xrightarrow{G'\alpha S} G'F'S$$

and

$$GFS \xrightarrow{G\alpha S} GF'S \xrightarrow{\beta F'S} G'F'S.$$

In fact, the two are the same--so that we may refer to either overall transformation as $\beta\alpha$ --as we see by proving the commutativity of the outer square of the diagram:

$$\begin{array}{ccc} GFS & \xrightarrow{G\alpha S} & GF'S \\ {\scriptstyle \beta FS}\downarrow & B\alpha S & \downarrow{\scriptstyle \beta F'S} \\ G'FS & \xrightarrow{G'\alpha S} & G'F'S \end{array}$$

But this is immediate, for this diagram is simply that expressing that β is a natural transformation when applied to $FS \xrightarrow{\alpha S} F'S$. The usual calculation shows that <u>horizontal composition</u>, so defined, indeed yields a functor.

Further, property (1) is satisfied in a strengthened form, for composition is associative up to identity. It is then trivial that the identity endofunctor $1 = id_{\mathcal{A}}$ plays for $\mathcal{A}^{\mathcal{A}}$ the role that 1 plays for $\mathbf{Set}$

$$1F = F = F1$$

Note that $\mathbf{Set} \times \mathbf{Set} \xrightarrow{\times} \mathbf{Set}$ does indeed become $\mathbf{Set}^{\mathbf{Set}} \times \mathbf{Set}^{\mathbf{Set}} \xrightarrow{comp} \mathbf{Set}^{\mathbf{Set}}$ under the embedding $X \to \circ \times X$:

$$(\circ \times X)(\circ \times Y) \cong \circ \times (X \times Y)$$

and we do not distinguish S from $S \times 1$ so that $\circ \times 1 \cong id_{\mathbf{Set}}$. Under the embedding which sends the set M to a functor T, $M \times M$ goes to the composition of T with itself, and $M \times M \times M \mapsto TTT$; the map $M \times M \xrightarrow{m} M$ goes to a natural transformation $TT \xrightarrow{\mu} T$, while $1 \xrightarrow{e} T$ yields a natural transformation $1 \xrightarrow{\eta} T$. Thus our definition of a monoid in $\mathbf{Set}$ takes the form of a <u>monad</u> (also known as a <u>triple</u>) where:

5.3 DEFINITION: $\mathbb{T} = (T,\eta,\mu)$ is a monad in the category $\mathcal{A}$ with unit η and multiplication μ if $\mathcal{A} \xrightarrow{T} \mathcal{A}$ is a functor and if $1 \xrightarrow{\eta} T$ and $TT \xrightarrow{\mu} T$ are natural transformations subject to the axioms:

identity axioms:

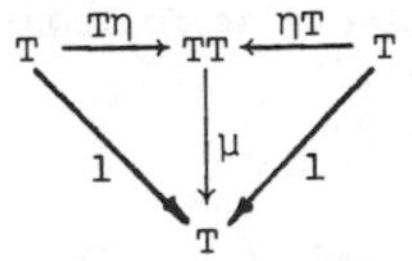

associativity axiom:

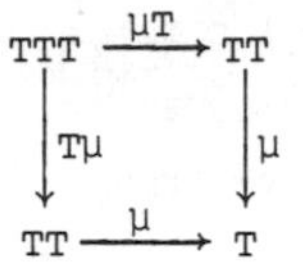

Just as a monoid homomorphism is a map $M \xrightarrow{\lambda} M'$ such that $\lambda(e) = \lambda(e')$ while $(\lambda\times\lambda)m' = (m\times m)\lambda$, so do we have:

5.4 DEFINITION: A monad homomorphism $\mathbb{T} \to \mathbb{S}$ is a natural transformation $\mu : T \xrightarrow{\bullet} S$ which satisfies the commutativity of

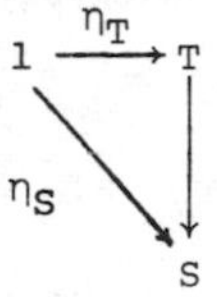

and

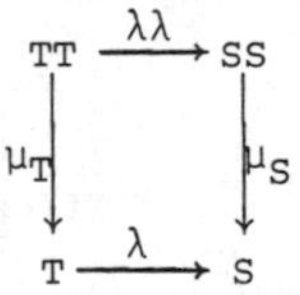

If X_0 is an alphabet, the monad associated with the free monoid $X_0{}^*$ is defined by

$$T = -\times X_0{}^*$$

$$\mu : -\times X_0{}^*\times X_0{}^* \xrightarrow{\bullet} -\times X_0{}^* \qquad \eta : I \xrightarrow{\bullet} -\times X_0{}^*$$

$$(-,w,w') \mapsto (-,ww') \qquad - \mapsto (-,\Lambda)$$

where Λ is the empty word. Clearly, the above definitions satisfy the axioms. Furthermore, if $X_0{}^*$ is replaced by an arbitrary monoid S, not necessarily free, it is clear how a monad can be formed this way.

The theory of tree processing yields another example of a monad. Let Σ be a ranked alphabet, i.e., a set Σ together with a finite relation $r \subset \Sigma\times N$ called the ranking relation. If $r(\sigma,n)$ we say σ has rank n and we denote $\Sigma_n = \{\sigma : r(\sigma,n)\}$ (r may be identified with the finitary operator domain (see part 3) $\Omega_n = \Sigma_n\times\{n\}$; in the ranking relation point of view the same symbol may have different arities). Let Z be a set of variables. The set $T_{\Sigma,Z}$ of finite Σ-trees on Z generators is a subset of $\{\Sigma \{[,]\} Z'\}^*$, where $Z' = \{<z> : z \varepsilon Z\}$ and we assume that $\{[,]\}$, $\{<,>\}$, Z and Σ are pairwise disjoint.

$T_{\Sigma,Z}$ is defined inductively as follows:

$$Z' \cup \Sigma_0 \subset T_{\Sigma,Z}$$

$$\sigma \varepsilon \Sigma_n \text{ and } t_1,\ldots,t_n \varepsilon T_{\Sigma,Z} \Rightarrow$$

$$\sigma[t_1,\ldots,t_n] \varepsilon T_{\Sigma,Z}.$$

$\sigma[t_1,\ldots,t_n]$ is interpreted as the tree

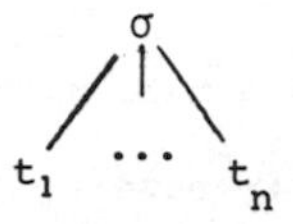

Define a functor T : Set $\rightarrow$ Set to have the object function $TZ = T_{\Sigma,Z}$ (i.e., T sends a set Z to the set of finite Σ-trees on Z generators) and the mapping function defined inductively by:

$$f : Z \rightarrow U$$

$$Tf(<z>) = <f(z)>$$

$$Tf(\sigma[t_1,\ldots,t_n]) = \sigma[Tf(t_1),\ldots,Tf(t_n)]$$

i.e., Tf simply relabels each $z \varepsilon Z$ with $f(z) \varepsilon U$. Here we denote $<z> \varepsilon T_{\Sigma,Z}$ as a tree, to distinguish it from the variable $z \varepsilon Z$.

The tree monad (which is the same as the free monad $(X_\Omega)^@$ of part 3) over Σ consists of the above functor T (called the tree functor) and the natural transformations η and μ where $\eta : I \dot{\rightarrow} T$ is the inclusion of generators, defined on elements as

$$\eta z : z \mapsto <z>$$

TTZ is a set of Σ-trees on generators which are Σ-trees on Z generators. $\mu Z : TTZ \rightarrow TZ$ removes one (outer) level of brackets, so that, for example, if

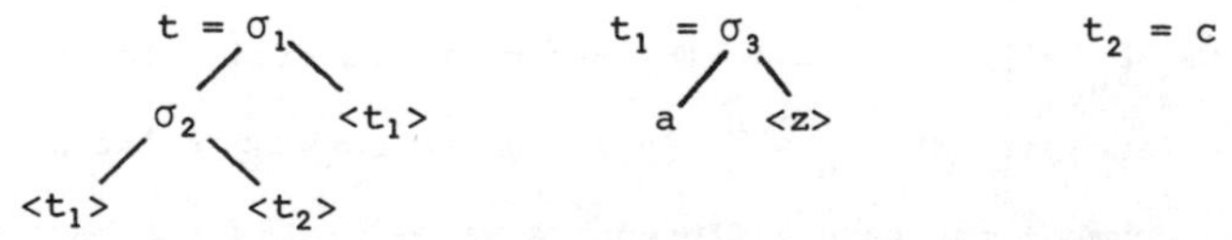

then

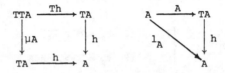

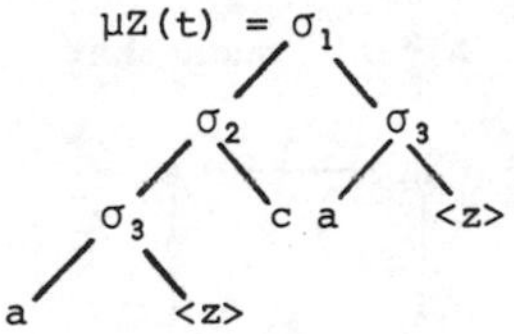

<u>5.5 DEFINITION</u>: (i) If (T,η,μ) is a monad in $\mathcal{K}$, a <u>T-algebra</u> (A,h) is a pair consisting of an object A of $\mathcal{K}$ (the underlying object, or <u>carrier</u>, of the algebra) and a morphism $h : TA \to A$ of $\mathcal{K}$ (the <u>structure map</u> of the algebra) such that

$$\begin{array}{ccc} TTA & \xrightarrow{Th} & TA \\ \downarrow{\scriptstyle \mu A} & & \downarrow{\scriptstyle h} \\ TA & \xrightarrow{h} & A \end{array} \qquad \begin{array}{ccc} A & \xrightarrow{A} & TA \\ & {\scriptstyle 1_A}\searrow & \downarrow{\scriptstyle h} \\ & & A \end{array}$$

The first diagram is called the <u>associativity</u> axiom and the second the <u>unitary</u> axiom.

(ii) A <u>morphism</u> $f : (A,h) \to (A',h')$ <u>of T-algebras</u> is a morphism $f : A \to A'$ of which renders commutative the diagram

$$\begin{array}{ccc} TA & \xrightarrow{h} & A \\ \downarrow{\scriptstyle Tf} & & \downarrow{\scriptstyle f} \\ TA' & \xrightarrow{h'} & A' \end{array}$$

With the above definitions T-algebras and their morphisms constitute a category, denote it $\mathcal{K}^T$.

A Σ-algebra is a pair (A,δ) where A is a set, called the <u>carrier</u> of the algebra, $\delta : \Sigma A \to A$ is a function and Σ is a ranked alphabet which can be interpreted as a functor <u>Set</u> $\to$ <u>Set</u> (cf. X_Ω in part 3) defined on objects as

$$\Sigma A = \{\sigma[\langle a_1\rangle,\ldots,\langle a_n\rangle] \ : \ \sigma\varepsilon\Sigma_n \ , \ r(\sigma,n) \ \text{ and } \ a_1,\ldots,a_n\varepsilon A\}.$$

For $f : A \to A'$ set

$$\Sigma f \ : \ \Sigma A \to \Sigma A'$$

$$: \ \sigma[\langle a_1\rangle,\ldots,\langle a_n\rangle] \mapsto \sigma[\langle f(a_1)\rangle,\ldots,\langle f(a_n)\rangle]$$

Functoriality is easy to check. A homomorphism $f : (A,\delta) \to (A',\delta')$ of Σ-algebras is a function $f : A \to A'$ such that

$$\begin{array}{ccc} \Sigma A & \xrightarrow{\delta} & A \\ \downarrow{\scriptstyle \Sigma f} & & \downarrow{\scriptstyle f} \\ \Sigma A' & \xrightarrow{\delta'} & A' \end{array}$$

For $\sigma\varepsilon\Sigma_0$ we write δ_σ for $\delta(\sigma[\Lambda])$. Σ-algebras and their homomorphisms constitute a category, denoted Σ-alg.

5.6 PROPOSITION: Let T be the tree functor over Σ and (A,δ) a Σ-algebra. Define $h : T_{\Sigma,A} \to A$ by

$$h(t_0) = \begin{cases} \delta_\sigma & \text{if } t_0 = \sigma \in \Sigma_0 \\ a & \text{if } t_0 = \langle a \rangle, \ a \varepsilon A \end{cases}$$

$$h(\sigma[t_1,\ldots,t_n]) = \delta(\sigma[\langle h(t_1)\rangle,\ldots,\langle h(t_n)\rangle])$$

Then the pair (A,h) is a T-algebra.

5.7 THEOREM: An adjunction $(F,U,\eta,\varepsilon) : \mathcal{A} \to \mathcal{B}$ determines a monad $(UF,\eta,U\varepsilon F)$ in the category $\mathcal{A}$.

Proof: From the triangular identities , putting F behind in the first diagram and U in front in the second one we get the unitary axioms for the monad $(UF,\eta,U\varepsilon F)$. The diagram

$$\begin{array}{ccc} FUFU & \xrightarrow{FU\varepsilon} & FU \\ \downarrow{\scriptstyle \varepsilon FU} & & \downarrow{\scriptstyle \varepsilon} \\ FU & \xrightarrow{\varepsilon} & I \end{array}$$

is just the definition of horizontal composition and commutes because ε is a natural transformation. Putting U in front and F behind in the above diagram we get

$$\begin{array}{ccc} (UF)(UF)(UF) & \xrightarrow{UF(U\varepsilon F)} & (UF)(UF) \\ \downarrow{\scriptstyle (U\varepsilon F)UF} & & \downarrow{\scriptstyle U\varepsilon F} \\ (UF)(UF) & \xrightarrow{U\varepsilon F} & (UF) \end{array}$$

the associativity axiom for the monad $(UF,\eta,U\varepsilon F)$. □

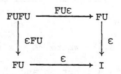

5.8 THEOREM: If (T,μ,η) is a monad in $\mathcal{K}$, then the class of all T-algebras and their morphisms form the category $\mathcal{K}^T$. There exists an adjunction

$$\langle F^T, U^T, \eta^T, \varepsilon^T\rangle : \mathcal{K} \rightharpoonup \mathcal{K}^T$$

in which the functors U^T and F^T are given by the respective assignments

$$U^T: \quad \begin{array}{c} (A,h) \\ f\big\downarrow \\ (A',h') \end{array} \quad \mapsto \quad \begin{array}{c} A \\ \big\downarrow f \\ A' \end{array}$$

$$F^T: \quad \begin{array}{c} A \\ f\big\downarrow \\ A' \end{array} \quad \mapsto \quad \begin{array}{c} (TA,\mu A) \\ Tf\big\downarrow \\ (TA',\mu A') \end{array}$$

while $\eta^T = \eta$, and $\varepsilon^T(A,h) = h$ for each T-algebra (A,h). The monad defined in $\mathcal{K}$ by this adjunction is the given monad (T,μ,η).

5.9 THEOREM: Let (T,μ,η) be a monad in the category $\mathcal{K}$. Associate to each object A in $\mathcal{K}$ a new object A_T, and to each morphism $f : A \to TB$ in $\mathcal{K}$ a new morphism $f^b : A_T \to B_T$. These new objects and morphisms form a category $\mathcal{K}_T$, the Kleisli category of T, when we define composition by

$$A_T \xrightarrow{f^b} B_T \xrightarrow{g^b} C_T = (\mu C \cdot Tg \cdot f)^b$$

The identity for A_T is $(\eta A)^b$.

Moreover, functors $F_T : \mathcal{K} \to \mathcal{K}_T$ and $U_T : \mathcal{K}_T \to \mathcal{K}$ are defined by

$$F_T: \quad k : A \to B \mapsto (\eta B \cdot k)^b : A_T \to B_T$$

$$U_T: \quad f^b : A_T \to B_T \mapsto \mu B \cdot Tf : TA \to TB.$$

The bijection $f^b \mapsto f$ yields the adjunction $\mathcal{K} \underset{U_T}{\overset{F_T}{\rightleftarrows}} \mathcal{K}_T$ which defines in $\mathcal{K}$ precisely the given monad (T,μ,η). □

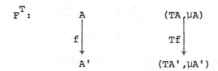

6. CLOSED AND MONOIDAL CATEGORIES

REMARK: (i) If $\mathcal{K} \to \underline{1}$ and $\Delta : \mathcal{K} \to \mathcal{K} \times \mathcal{K}$ have left adjoints, then $\mathcal{K}$ has an initial object and finite coproducts.

(ii) If $\mathcal{K} \to \underline{1}$ and $\Delta : \mathcal{K} \to \mathcal{K} \times \mathcal{K}$ have right adjoints, then $\mathcal{K}$ has a terminal object and finite products.

Thus all axioms in the following definition ask for the existence of adjoints.

6.1 DEFINITION: The category $\mathcal{C}$ is called Cartesian closed if it is equipped with a terminal object; a bifunctor $\mathcal{C} \times \mathcal{C} \to \mathcal{C}$ which assigns a product to each pair of objects; and a right adjoint $(-)^B$ to each $-\times B$ for B in $\mathcal{C}$. We call the $\varepsilon_C : C \leftarrow C^B \times B$ which yields

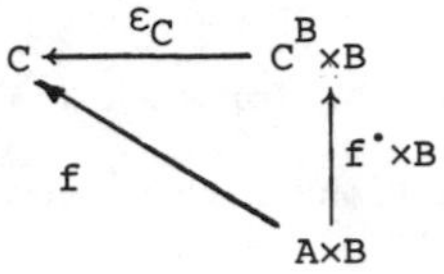

the evaluation map.

Set is cartesian closed, with $C^B = \underline{\text{Set}}(B,C)$ In R-Mod, we take $\otimes_R$ to be the tensor product, and find that $-\otimes_R B$ has right adjoint R-Mod[B,-]. This is not a cartesian closed category, but is closed in the sense of Definition 6.9 , since $\otimes_R$ is not the categorical product

The considerations of Section 5 suggest the following general definitions:

6.2 DEFINITION: A monoidal category* $(\mathcal{C},\otimes,I)$ comprises a category $\mathcal{C}$, a functor $\mathcal{C} \times \mathcal{C} \xrightarrow{\otimes} \mathcal{C}$ and an object $I \in \mathcal{C}$ such that

1. "I is a 2-sided unit for ⊗": There are natural equivalences ℓ, r of functors $\mathcal{C} \to \mathcal{C}$ such that

$$I \otimes A \xrightarrow[\cong]{\ell_A} A \quad \text{and} \quad A \otimes I \xrightarrow[\cong]{r_A} A$$

*See Mac Lane for a further pentagonal diagram, often imposed, which is needed to prove that "all diagrams commute up to isomorphism".

2. "⊗ is associative": There is a natural equivalence a of functors $\mathcal{C}\times\mathcal{C}\times\mathcal{C} \to \mathcal{C}$ such that

$$(A\otimes B) \otimes C \xrightarrow[\cong]{a_{A,B,C}} A \otimes (B\otimes C)$$

We say $(\mathcal{C},\otimes,I)$ is strictly monoidal if, further, a, ℓ and r are equalities.

6.3 DEFINITION: A monoid in $(\mathcal{C},\otimes,I)$ is then (M,m,e) where M is an object of $\mathcal{C}$, while $M\otimes M \xrightarrow{m} M$ and $I \xrightarrow{e} M$ are $\mathcal{C}$-morphisms such that "m is associative" and "e is a 2-sided unit" in the sense specified by commutativity of the following diagrams:

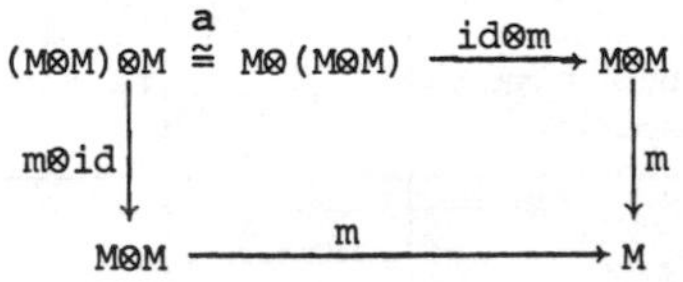

and

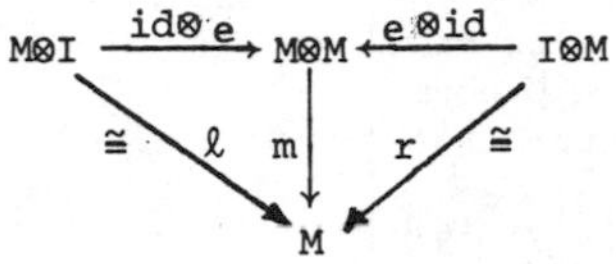

Note that the category Cat of categories (with functors for morphisms) underlies a monoidal category $(\underline{cat},\times,1)$ where 1 is the terminal (one-morphism) category. A strictly monoidal category is then simply a $(\underline{cat},\times,1)$-monoid.

6.4 EXAMPLES:

	⊗	I	Monoid
(sets)	×	1 one element	ordinary monoid
Groups	×	{e}	abelian group This is because a group G with multiplication • and identity 1 only supports a monoid (G,m,c) if $m(a_1,a_2) = m((a_1,1)(1,a_2))$ $= m(a_1,1)m(1,a_2)$ $= a_1a_2$ $= m((1,a_2)(a_1,1))$ $= a_2a_1$ and so G must be abelian.

	$\otimes$	I	Monoid
Abelian Groups	$\otimes_{\underline{\underline{Z}}}$	$\underline{\underline{Z}}$	ring
Abelian Monoids	$\otimes$	$\underline{\underline{N}}$	semiring
R-Mod (R commutative ring with 1)	$\otimes_R$ $\otimes$ provides the universal property of linearizing bilinear maps.	R	R-algebra
Topological Spaces	$\times$	1	topological monoid (m jointly continuous)
	$\otimes$ provides the universal property of making separately continuous maps continuous.	1	topological monoid (m separately continuous)
Banach Spaces with Norm-Decreasing Maps	$\otimes$ Schatten tensor product: Pick the largest $\|\cdot\|$ norm on $X\otimes Y$ such that $\|x\otimes y\| \leq \|x\| \cdot \|y\|$ and take the completion. $\otimes$ is then the left adjoint of Hom (cf. Schatten's monograph).	scalar field (real or complex)	Banach Algebra

As we have already seen in Section 5, the most natural example of a strictly-monoidal category is of interest to automata theorists:

6.5 THEOREM: For any category $\mathcal{K}$, the category $\mathcal{K}^{\mathcal{K}}$ of endofunctors [i.e., (not necessarily input) processes] is a strictly monoidal category, with $\otimes$ = horizontal composition and $1 = id_{\mathcal{K}}$. □

Again, the definition of monad in Section 5 tells us that a $\mathcal{K}^{\mathcal{K}}$ - monoid is just a monad $\mathbb{T} = (T,\eta,\mu)$. Although we shall not use it, it is worth recalling here the relationship with universal algebras (which may certainly be infinitary) when $\mathcal{K}$ = Set:

6.6 THEOREM: Any category of (Ω,E)-algebras is isomorphic to Set for some monad $\mathbb{T}$ with rank [i.e., there exists a cardinal α such that whenever S is a set and $\bar{s}\in TS$ then there exists $n\in\alpha$ and $f:\alpha\to S$ with $\bar{s}$ in the image of Tf] and $\mathbb{T}$ is unique up to isomorphism; conversely, every monad $\mathbb{T}$ with rank admits a presentation Set $\cong$ (Ω,E)-algebras for some (Ω,E). □

We generalize the notion of monoid homomorphisms to obtain:

6.7 DEFINITION: Given a monoidal category $(\mathcal{C},\otimes,I)$ we define $\underline{\mathrm{Mon}}(\mathcal{C})$ to be the category of $(\mathcal{C},\otimes,I)$-monoids, where $M \xrightarrow{f} M'$ is a monoid homomorphism $(M,m,e) \xrightarrow{f} (M',m',e')$ just in case

$$\begin{array}{ccc} M\otimes M & \xrightarrow{f\otimes f} & M'\otimes M' \\ {\scriptstyle m}\downarrow & & \downarrow{\scriptstyle m'} \\ M & \xrightarrow{f} & M \end{array} \qquad \begin{array}{ccc} & I & \\ {\scriptstyle e}\swarrow & & \searrow{\scriptstyle e'} \\ M & \xrightarrow{f} & M' \end{array}$$

6.8 DEFINITION: A monoidal category is symmetric when it is equipped with isomorphisms

$$\gamma_{A,B} : A\otimes B \cong B\otimes A$$

natural in A and B, such that the diagrams

$$\gamma_{A,B}\circ\gamma_{B,A} = 1, \qquad r_B = \ell_B\circ\gamma_{B,I} : B\otimes I \cong B$$

and

$$\begin{array}{ccccc} A\otimes(B\otimes C) & \xrightarrow{a} & (A\otimes B)\otimes C & \xrightarrow{\gamma} & C\otimes(A\otimes B) \\ {\scriptstyle 1\otimes\gamma}\downarrow & & & & \downarrow{\scriptstyle a} \\ A\otimes(C\otimes B) & \xrightarrow{a} & (A\otimes C)\otimes B & \xrightarrow{\gamma\otimes 1} & (C\otimes A)\otimes B \end{array}$$

all commute. [This yields the validity of "all" such interchanges.]

6.9 DEFINITION: A symmetric monoidal category $\mathcal{K}$ is a closed category if each functor $-\otimes B : \mathcal{K}\to\mathcal{K}$ has a specified right adjoint $(\ )^B : \mathcal{K}\to\mathcal{K}$. All cartesian closed categories, such as Set, are closed categories. In all these cases, the functor $(\ )^B : \mathcal{K}\to\mathcal{K}$ is a sort of "internal hom functor".

6.10 DEFINITION: A *topos* is a cartesian-closed category together with an object Ω and a map $1 \xrightarrow{t} \Omega$ such that for any monomorphism $A' \to A$ there exists a unique $A \to \Omega$ such that

$$\begin{array}{ccc} A' & \longrightarrow & A \\ \downarrow & & \downarrow \\ 1 & \xrightarrow{t} & \Omega \end{array}$$

is a pullback.

The immediate interest of topos is given by

6.11 PROPOSITION: For any small category $\mathcal{K}$, the functor category $\underline{\underline{\mathrm{Set}}}^{\mathcal{K}}$ is a topos.

Topos theory, an exciting new application of category theory, provides a common framework for intuitionistic models of set theory and sheaf-categories arising in algebraic geometry. See [Toposes, Algebraic Geometry & Logic, Springer Lecture Notes, 274] and references there. About one-half of the talks at the week-long summer 1973 Tagung uber Kategorien at Oberwolfach, Germany, were about topoi. Scott's cartesian closed category of continuous lattices is, alas, not a topos.

A CONTROL THEORIST LOOKS AT ABSTRACT NONSENSE

B. D. O. Anderson
Electrical Engineering
University of Newcastle
NSW, Australia

The purpose of this part of the introduction is to give those at the category theory --rather than the control-- end of the spectrum some feel for the physical origin of some control problems, and some feel for some of the viewpoints taken by control theorists of these problems.

In the space available, far more must be omitted than can be included and the treatment must necessarily be very superficial. These facts notwithstanding, the material is of vital significance to any category theorist who wishes to make an honest, objective claim of applying category theory to control. In order that such category theorists might have their work accepted by the control fraternity, it is essential that the control fraternity be convinced that the category theory is not category theory for its own sake, nor another way of viewing known control results, nor a way of getting category theory generalizations of known control results with the generalizations possessing no control theoretic significance. There are already several distinct ways of viewing many control results, and control theorists will be understandably reluctant to add a further way if it provides merely a new view, rather than new control results.

Perhaps category theorists could even be warned that the module-theoretic approach to linear systems initiated nearly ten years ago by R. E. Kalman, though undoubtedly ingenious and aesthetically appealing to the more mathematically oriented of the control fraternity, has not yet received the close attention of a great many control theorists. This is certainly no reflection on the intrinsic merit of the work; it is the result of control theorists saying --not necessarily with irrefutable logic-- "this theory tells me all about realization and pole positioning, albeit in a nice way, but these are things I know about. There are many other control problems, solved and unsolved, which it does not tell me about. Why should I bother with it?". Perhaps the real point is that the control theorists simply should work harder. Nevertheless, with this experience, it is a little disappointing to find so much attention being paid in this conference to the realization problem, to the exclusion of other problems. A control theorist could validly ask "When is category theory going to start dealing with other control issues, such as feedback?".

PHYSICAL ORIGINS FOR LINEAR SYSTEMS

One readily understood physical situation which led electrical engineers to start the train of thought, the current development of which is explored in the present volume, is exemplified by a network of inductors, resistors and capacitors, with a terminal pair at which a voltage may be applied and a resultant current measured.

Here one has a situation which can be described in a fairly obvious manner by a differential equation of the type (see [1])

$$\sum_{i=0}^{n} a_i \, y^{(i)} = \sum_{j=0}^{m} b_j \, a^{(j)} \qquad (1)$$

(u = voltage or input or control, y = current or output, a_i, b_j scalar real constants). One can replace this equation set, using a standard device, by one of the form

$$\dot{x} = Fx + gu \qquad y = hx \qquad (2)$$

where $x = [y \ \dot{y} \ \cdots \ y^{(n-1)}]$, F is an n x n real constant matrix and g is a real constant n-vector. In fact, an equivalent description is provided by

$$\dot{\bar{x}} = \bar{F}\bar{x} + \bar{g}u \qquad y = \bar{h}\bar{x} \tag{3}$$

where $\bar{x} = Tx$ for some nonsingular T, $\bar{F} = TFT^{-1}$, $\bar{g} = Tg$, and $\bar{h} = [1 \ 0 \ \cdots \ 0] \ T^{-1}$. Parenthetically, we comment that this sort of change of variable (via change of coordinate basis) is widely used in linear systems theory. For an excellent introduction to this subject, consult [2] . For a deeper discussion of a subset of the topics covered in [2] , see [3] .

Electrical engineers would think of (1) as providing an input-output description of the network, and (2) or (3) as providing an internal description (because of the presence of the intervening variable x, linking u and y. It can be, as a result of special choice of T, that x has some physical significance as far as the internal behaviour of the network is concerned; for example, the entries of x may correspond to the capacitor voltages and inductor currents within the network (see [1] and also [4] .)

It is clearly of interest to be able to compute the output at a given time resulting from the input applied up to that time, assuming the network is initially unexcited. In fact, one has

$$y(0) = \int_{-\infty}^{0} w(\tau)u(\tau)d\tau \tag{4}$$

(provided that the integral exists), where $w(\tau)$ is given by $\bar{h}\exp(-\bar{F}\tau)\bar{g}$. Thus there is defined a map $u(\tau), \ \tau \in (-\infty,0] \longmapsto y(0)$. Such a map is also an example of an input-output description ([1,2]), and the function $w(\cdot)$ is almost the <u>impulse response</u> of linear systems theory; here, (4) is almost the <u>convolution</u> formula ([1,2]).

Generally, from a system user's point of view, the input-output description is the most relevant. From a system designer's point of view, an internal description may be relevant, since generally it is needed in specifying how a system may physically be constructed.

Passing from an input-output description to an internal description is the act of solving the realization problem. Questions such as the following occur:

1) What $w(\tau)$ can be expressed in the form $w(\tau) = h \exp(-F\tau)g$?

2) How may one find any triple F, g, h realizing $w(\tau)$?

3) Are there especially interesting F, g, h realizing $w(\tau)$?

In realization theory, for those $w(\tau)$ expressible as $h \exp(-F\tau)g$, "especially interesting" has come to mean "with F matrix of minimum dimension", and the associated triples (F,g,h) are termed minimal or canonical (see [2,3]).

Minimal realizations are precisely those which are simultaneously completely reachable and completely observable. A realization is completely controllable if, with x initially zero, the map $u(\tau)$, $\tau \in (-\infty,0]$ $\longmapsto$ $x(0)$ defined by (3) -- well defined if, for example, u has compact support, and $x(-M)$ is zero where supp $u \in [-M,0]$ -- is onto; in other words, there exists a control taking the zero state to any nonzero state. A realization is completely observable if, with $u(\tau) = 0$ for $\tau \geq 0$, the map $x(0) \longmapsto y(\tau)$, $\tau \in [0,\infty)$ is injective; in other words, no two distinct states can yield the same output or, in theory (and actually in practice), the state $x(0)$ is computable from the output $y(\tau)$, $\tau \in [0,\infty)$.

The controllability property has many equivalent statements. For example ([2,3])

a) rank $[g \; Fg \; \cdots \; F^{n-1}g] = n$ (which is a helpful statement for checking controllability)

b) $w'e^{Ft}g = 0$ for all t implies $w = 0$

c) the eigenvalues of $F + gk'$ can be assigned arbitrarily via choice of k, provided that complex eigenvalues occur in complex conjugate pairs (this result is of major significance in the design of feedback control laws which are mentioned further below).

Likewise, the observability property has many equivalent statements. To begin with, it is dual to the controllability property. To fully expound the duality would not be worthwhile here; we merely comment that $[F,h]$ is a completely observable pair if and only if its adjoint pair $[F',h']$ is completely controllable. A second important property of observable systems is that one can construct a dynamic observer, the output of which is a vector $\hat{x}$, the input of which is both $u(\cdot)$ and $y(\cdot)$, such that $\lim_{t\to\infty} \hat{x}(t) - x(t) = 0$, with the convergence being exponential, and otherwise arbitrarily fast. The observer is actually itself a linear finite-dimensional system. (For some discussion of the principal types of observer, developed by Luenberger and Kalman-Bucy, see [5]).

Of course, not all the controllability equivalences carry over to more general category-theoretic situations; thus one might expect to carry over the epi property of controllability, but not the eigenvalue positioning property.

Let us pause to note a key notion of control--that of feedback. Consider the equation (3) in which we have said u is an input or control. Suppose u is derived as the sum of an externally applied control, v say, and a linear function $k'\bar{x}$ (k a real vector) of the state $\bar{x}$. Then we can say that there is feedback around the system defined by (3) which can be written as

$$\dot{\bar{x}} = \bar{F}\bar{x} + \bar{g}(v + k'\bar{x}) = (\bar{F} + \bar{g}\,k')\bar{x} + \bar{g}v$$

Feedback can be used to modify the style of behaviour of a system and it need not be linear as above. The principal design task in many control system problems is that of selecting a feedback law to cause the system with feedback --the closed loop system-- to have a certain type of behaviour; so feedback is a very central concept --if not the central concept-- in control.

VARIATIONS OF THE BASIC SYSTEM EQUATIONS

Let us note other physical situations which can be modeled by variations on (1) - (4).

Discrete-time systems (see e.g. [3]). As discussed in part 3 of this introduction, it is possible to derive discrete-time equations of the form

$$x_{k+1} = Ax_k + bu_k$$
$$y_k = cu_k \tag{5}$$

from, say, (3). (Category theorists should be warned that there are prejudices within the control theory community as to whether one should use F, g and h or A, b and c in linear system equations. In the author's view, the arguments favoring either one over the other have little or no objective validity.) If one attempts to solve (3) on a digital computer, some time sampling is obviously necessary (strictly, level quantization is also needed, but this is overlooked). Sampling (3) leads to a set-up describeable by (5); u_k, y_k and x_k are sample values of u, y, x roughly, or averages over a sampling interval of the continuous time quantities. A, b, c can be found from F, g, h.

Non linear systems (see e.g. [6,7]). Many physical systems are nonlinear. The result is that one may have to work with equations such as

$$\dot{x} = f(x) + g(x)\,u \tag{6}$$

or, more generally,

$$\dot{x} = f(x,u) \tag{7}$$

and

$$y = h(x) \tag{8}$$

Linear time-varying systems. There are at least two quite distinct ways in which one might end up studying a time-varying system, even a discrete time-varying system. Some systems important in electrical engineering applications are intentionally made time-varying, for example, parametric amplifiers, where one has periodically varying F, g and/or h replacing the constant matrices of earlier (see [8]). Sampling such systems yields a time-varying discrete-time system.

A second source for such systems arises when one studies a perturbation of a nonlinear system about a nominal control and trajectory. Let a control $\bar{u}(\cdot)$ (the nominal control) give rise to a trajectory $\bar{x}(\cdot)$ via $\dot{\bar{x}} = f(\bar{x},\bar{u})$. Suppose an actual control $u(t) = \bar{u}(t) + \delta u(t)$ is applied, where δu is small. If the resultant trajectory is $x = \bar{x} + \delta x$ with small δx, then δx and δu may be approximately related by a time-varying differential equation, $(\dot{\delta x}) = F(t)(\delta x) + g(t)(\delta u)$. The entries of $F(\cdot)$ and $g(\cdot)$ are obtained by evaluating entries of partial derivatives of f, which vary along the trajectory. This sort of device is important in optimal control, see e.g. [3,9].

Infinite dimensional systems. Systems such as telephone channels, waveguides ([10]) and reactors are examples of infinite dimensional systems of great interest to engineers. Many such systems are linear and have external descriptions via a $w(\tau)$ as in (4), but do not have a finite-dimensional description such as (3). Since the engineering design problem for such systems clearly exists (given an input-output specification, build a system), and since system descriptions of a level adequate to build a system are generally of the internal description type (whatever that may be in this case), we see that there is some sort of realization problem here too.

Vector generalizations (see [1,2,3,4,5,6]). It is always possible to conceive of a number of input channels and a number of output channels. The generalization is not always trivial, even for linear, finite-dimensional systems.

The major tool through the 1960's and early 1970's in linear system theory and, to an extent, control theory generally, has been linear algebra. Perhaps because engineers on the whole are more comfortable in less abstract situations, or perhaps simply because engineering problems demand solutions specified by numbers, matrices rather than linear transformations figure largely. References containing many of the standard devices (or, as category theorists might prefer to say, tricks) are [11, 12]. Recently, [13] has appeared; it would provide a nonspecialist with the matrix theory scenario for linear systems.

Let us now discuss some of these ideas in a little more detail.

LINEAR DISCRETE-TIME SYSTEMS

One can start by postulating that a system is of the form

$$\begin{aligned} x_{k+1} &= Ax_k + Bu_k \\ y_k &= Cx_k \end{aligned} \tag{9}$$

where A, B, C are real constant matrices of dimension n x n, n x p, m x n respectively. Imagine p input channels; at time k, they inject u_k into the system. The state of the system at time k is x_k; the output is y_k.

One can conclude that

$$y_1 = \sum_{i=0}^{M} CA^i Bu_{-i} \tag{10}$$

where one assumes that the input sequence is of support bounded on the left, and the state is zero prior to nonzero input appearing. See [3].

Now one dresses this up a bit and observes that (10) is but a map from $R^m[z]$, the set of m-vectors whose entries are polynomials in z with real coefficients into the reals, if we recognize that the sequence $(0 \cdots 0\ u_{-M}\ u_{-(M-1)} \cdots u_0)$ is isomorphic to the polynomial

$$\sum_{i=0}^{M} z^i\, u_{-i}$$

Call this map $f : R^m[z] \longrightarrow R$.

Just as $R^m[z]$ is introduced to help talk about input sequences, we can introduce $R^p[[z^{-1}]]$, formal power series in z^{-1} with p-vector coefficients, to talk about output sequences.

Suppose an input sequence $(\cdots 0\ u_{-m}\ u_{-m+1} \cdots u_0\ 00 \cdots 0 \cdots)$ is applied. Look at the output string $(y_1, y_2, y_3 \cdots)$ (disregard all outputs earlier than time instant 1). The output strings are clearly isomorphic to $R^p[[z^{-1}]]$: $(y_1, y_2, y_3 \cdots) \longmapsto \sum y_i z^{-i}$.

Now it is clear that $f : (0 \cdots 0\ u_{-m} \cdots u_0) \longmapsto y_1$ can be extended to $f^* : (0 \cdots 0\ u_{-m} \cdots u_0) \longmapsto (y_1, y_2, y_3 \cdots)$ by direct calculation using (9). But there are two other distinct ways of looking at this extension.

First, as first pointed out by Kalman (see e.g. [3]), f^* <u>is an</u> $R[z]$ -<u>module</u> <u>homomorphism</u>. (That $R^m[z]$ is an $R[z]$ -module with appropriate module action is easily seen; that $R^p[[z^{-1}]]$ is also an $R[z]$ -module is a little harder to see: the action involves multiplication of each component by a polynomial in z and deletion of nonnegative powers of z).

Second, f^* is a morphism $(R^m)^{§} \longrightarrow (R^p)_{§}$ where $(R^m)^{§}$ is the copower (weak direct sum) of an infinite sequence of copies of R^m and $(R^p)_{§}$ is the product of an infinite sequences of copies of R^p, with a commutative diagram of the following sort holding

$$\begin{array}{ccc} (R^m)^{§} & \xrightarrow{\ z_u\ } & (R^m)^{§} \\ \big\downarrow f^* & & \big\downarrow f^* \\ (R^p)_{§} & \xrightarrow[\ z_y\ ]{} & (R^p)_{§} \end{array}$$

where z_u and z_y are morphisms defineable just using coproduct and product properties. See [14] .

Realization is the problem of passing from f to an (A,B,C) triple as in (9).

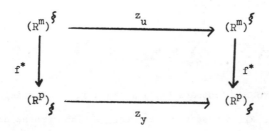

From the $R[z]$ -module viewpoint, one proceeds as follows. From f, set up f^*. Define the state-space X in which x resides by $X = R^m[z]/\ker f^*$. Then technical, but conceptually straightforward calculations allow the construction of an A,B,C triple, first as linear transformations, and then, via coordinate basis specialization, as matrices. By using standard module-decomposition theorems, one can get various particular structures for A, B and C. With this definition of X, realizations are guaranteed universal.

As an interesting sidelight to these calculations, one can observe that X, stripped of its module structure, is identical with the set of Nerode equivalence classes which may be associated with f. This is a pleasing result, but it would probably have been equally pleasing if instead, for example, the Nerode equivalence classes generated X.

To proceed using the categorical approach, one needs an $\mathcal{E}-\mathcal{M}$ factorization system (see the first part of this introduction). One sets $X = \mathrm{Im}(f^*)$. Then $A : X \longrightarrow X$ follows by a standard property of $\mathcal{E}-\mathcal{M}$ factorizations, $B = r\, in_0$ and $C = \pi_1 \sigma$ where $\sigma r = f^*$ is the $\mathcal{E}-\mathcal{M}$ factorization of f^*, in_0 is the coproduct injection corresponding to $u_0 \longmapsto (\cdots 0 \cdots 0 u_0)$ and π_1 is the product projection corresponding to $(y_1, y_2, y_3 \cdots) \longmapsto y_1$.

The maps f and f^* can be thought of as being defined by the collection of matrices M_i such that

$$y_1 = \sum_{i \geq 0} M_i u_{-i} \tag{11}$$

The realization problem (compare (10)) is then one of finding A,B,C such that $CA^iB = M_i$ for all nonnegative integers i. Interpretation of the procedure for realizing f^* can be given in terms of the M_i. See expecially [3].

EXTENSIONS USING MODULE THEORY IDEAS

There is no single decisively preeminent extension of the module theoretic approach to linear system realization. Rather, there are a number of examples; perhaps their multitude indicates the richness of the theory as a realization theory. It would seem however that the application of the ideas to control problems other than realization has to this point been minimal.

First, one can make the observation that the construction of the state-space X as $R^m[z]/\ker f^*$ does not use finite-dimensionality of the underlying linear system. Second, one can observe that the fact that R is the real field is inessential; realization over other fields is relevant to coding theory, e.g. [15]. Third, one can even replace R by a ring (an important result here is that if the ring is a Noetherian integral domain with identity, finite dimensional realizations exists if and only if they exists over the associated quotient ring, [16]). Such extensions, were they without applications, would be unappetizing to control theorists. Applications exist however; for example, Brockett and Willems ([17]) used the ring of circulant matrices to get efficient algorithms for solving problems arising from a class of linear partial differential equations. Johnson ([18]) shows that discretization of the one-dimensional heat equation yields a linear difference equation over $R^*[y]$, the quotient ring of $R[x,y]$ modulo the principal ideal generated by $xy - 1$.

EXTENSIONS USING CATEGORY THEORY IDEAS

In these Proceedings, the paper by Wyman ([19]) allows the ring k to be arbitrary, and replaces $k[z]$ by an arbitrary k-algebra of operators. The construction of f^* proceeds by an adjointness argument. Full development of the examples is unfortunately inhibited by length constraints. However, one class of examples is provided by those partial difference equations stemming from two-dimensional partial differential equations. Such equations arise in studying the processing of seismic signals --see recent issues of journals such as the <u>IEEE Transactions on Geoscience Electronics</u> and <u>IEEE Transactions on Audio and Electroacoustics</u>. However, one must still seriously ask whether, from the point of view of applications, the problem justifies the use of tools which most electrical engineers would regard as exotic.

Elsewhere, the paper of Arbib and Manes ([14]) shows that in formal terms, the realization problem can be viewed using coproducts, products and $\mathcal{E}-\mathcal{M}$ factorizations. It then follows that one can recover, for example, some of the group machine ideas of Brockett and Willsky ([20]). Finite-dimensionality is to be discussed in [21] . Further, the possibility of replacing the real field with an arbitrary ring becomes clear.

CONTINUOUS-TIME LINEAR SYSTEMS

Before discussing some of the approaches to the realization problem for continuous-time linear systems ([1,2,4,5,6]) let us write several facts distinguishing them from discrete-time linear systems.

1. $y = \dot{u}$ is the equation of a capacitor ([1]); if $u(\cdot)$ has a step discontinuity, but remains bounded, y becomes infinite.

2. One may seek to rule out infinite inputs and outputs; however, engineers find it useful to work with an input-output description of a linear system, viz. its <u>impulse</u> <u>response</u> (very closely related to $w(\tau)$ in (4), see [1,2]). The impulse response is useful because, from it, one can in theory obtain the output due any input. However, as the name suggests, it is the response which would result from a Schwartz impulse function ([22]), which is not finite.

3. $y = u$ is the equation of a resistor ([1]); y is only nonzero if u is. So one cannot look at the output arising only after the cessation of input, for it is identically zero. In other words, whereas in discrete time systems, one could study the relation between input sequences on $(-\infty,0]$ and outputs on $[1,\infty)$, one is really driven to study at the least continuous-time inputs on $(-\infty,0]$ and outputs on $[0,\infty)$. (Justification is also provided by, e.g., $y = \dot{u}$).

4. In a physical situation impulse functions or their derivatives cannot really exist. Models of the form $y = \dot{u}$ cease to be physically valid when either u or y becomes too large ([1]).

5. Let u(t), $t \in (-\infty, \infty)$ be the value of some system variable at time t. A physical measuring instrument can never measure u(t) precisely. Instead, it will measure something like

$$\int_{t-\epsilon}^{t+\epsilon} \varphi(\lambda)\, u(\lambda)\, d\lambda$$

for some function $\varphi(\cdot)$ and some number ϵ, both determined by the instrument.

The Schwartz distribution theory provides the answer to many of these problems ([22]). One can restrict attention for example to those situations in which all inputs and outputs are in $\mathcal{D}_+$, the set of functions with support bounded on the left and which are infinitely differentiable, see e.g. [23] . Or one can take the view that the inputs and possibly the outputs need not even be bounded, although they must be distributions (implying a measureability property in the sense of point 5 above).

The first attempt at a realization theory paralleling the module theoretic approach to discrete time linear systems appears to be that of Kalman and Hautus ([24]). They start by postulating that the system defines a linear map $f : \mathcal{E}'^{\,m}_{(-\infty,0]} \longrightarrow R$ (here $\mathcal{E}'$ is the space of distributions with compact support). Then f is extended to an $\mathcal{E}'_{(-\infty,0]}$ homomorphism $f^* : \mathcal{E}'^{\,m}_{(-\infty,0]} \longrightarrow \mathcal{E}^{p}_{[0,\infty)}$ (here, $\mathcal{E}$ denotes C^∞ functions). Then one looks at the Nerode equivalence classes, observing them to be isomorphic to $\mathcal{E}'^{\,m}_{(-\infty,0]}/\ker f^*$, which is an $\mathcal{E}'_{(-\infty,0]}$ homomorphic image. Finally, one can examine the consequences of finite-dimensionality.

There are difficulties with this approach. First, even such a system as $\dot{y} = u$ is not included. Second, though outputs are C^∞ and the states are equivalence classes of outputs, the states have to be viewed as being distributions.

More recently, Kamen ([25]) considered maps

$f^* : \mathcal{E}'^{\,m}_{(-\infty,0]} \longrightarrow \mathcal{D}'^{\,p}_{(0,\infty)}$ (Here $\mathcal{D}'$ is the space of distributions). Kamen abandons the attempt to start with a map f with codomain R in view of the difficulties that arise. However, his definition of f^* also fails to capture y = u or $y = \dot{u}$, both of interest to engineers, because the output interval is $(0,\infty)$ rather than $[0,\infty)$. One finds that f^* is an $\mathcal{E}'_{(-\infty,0]}$ -module homomorphism, and as a result, can obtain structure theorems for some finite-dimensional problems. These appear to be promising in connection with the study of lumped-distributed systems ([26]), such as those obtained by interconnection of a finite number of resistors, inductors, capacitors and transmission lines. In these Proceedings, Kamen ([27]) takes some of these ideas further. The paper largely speaks for itself, but there is one point we wish to make here. To get around the difficulty of overlapping input and output intervals in the definition of f, the technique adopted demands that concatenation of two inputs only be permitted with the inputs are separated by an interval of length a (a is arbitrary, > 0) during which the input is zero. This is slightly unfortunate, but perhaps less serious a restriction than those of earlier papers.

REFERENCES, WITH COMMENTS

As far as possible, only books have been included, in the belief that they are a more suitable starting point for the novice.

1. C. A. Desoer and E. S. Kuh, Basic Circuit Theory, McGraw-Hill, 1970.

(This text is valuable for various linear systems viewpoints, all within the framework of networks. The sorts of network descriptions used include ordinary differential equations like (1), state-variable systems like (2), impulse response description (related to (4)) and Laplace transform and Fourier transform descriptions. The latter two, though differing but trivially from the mathematical viewpoint in many cases, provide to an engineer vastly differing heuristically derived information.)

2. R. W. Brockett, Finite-Dimensional Linear Systems, John Wiley, 1970.

3. R. E. Kalman, P. L. Falb and M. A. Arbib, Topics in Mathematical System Theory, McGraw-Hill, 1969.

4. B. D. O. Anderson and S. Vongpanitlerd, Network Analysis and Synthesis: A Modern Systems Theory Approach, Prentice Hall, 1973.

(This text discusses network theory almost exclusively using state-variable ideas. It contains a one chapter survey of many results concerning finite-dimensional linear systems).

5. B. D. O. Anderson and J. B. Moore, Linear Optimal Control, Prentice-Hall, 1971.

(This book contains a discussion of observers, both in a deterministic framework and a stochastic framework --the latter because one needs to model and cope with the effects of measurement noise picked up by the sensors associated with a physical linear system.)

6. L. A. Zadeh and C. A. Desoer, Linear System Theory -A State Space Approach, McGraw-Hill, 1963.

(This book was the first to present system theory in a form that engineers could understand. It is a source of many valuable ideas.)

7. R. R. Mohler and A. Ruberti, eds., Theory and Application of Variable Structure Systems, Academic Press, 1972.

(This book is a collection of papers, some offering examples of nonlinear systems).

8. D. G. Tucker, Circuits with Periodically-varying Parameters, Unending Modulators and Parametric Amplifiers, Van Nostrand, 1964.

9. M. Athans and P. L. Falb, Optimal Control, McGraw-Hill, 1966.

(This book has a fairly lengthy introduction to linear systems theory before entering on a discussion of optimal control theory).

10. S. Ramo, J. R. Whinnery and T. van Duzen, Fields and Waves in Communication Electronics, Wiley, 1965.

11. F. R. Gantmakher, Theory of Matrices, 2 vols., Chelsea, 1960.

12. R. E. Bellman, Introduction to Matrix Analysis, McGraw-Hill, 1960.

13. S. Barnett, Matrices in Control Theory, Van Nostrand, 1971.

14. M. A. Arbib and E. G. Manes, Foundations of system theory: decomposable machines, Automatica, to appear.

(This is an interesting paper, providing a better unification of some results of linear systems and group machines then hitherto available. Whether the approach will catch on with the control fraternity or not may well depend on what the contents of sequel papers are).

15. J. L. Massey, Shift-register synthesis and BCH decoding, IEEE Trans. Inf. Th. IT-15, 1969, 122-127.

16. Y. Rouchaleau, B. F. Wyman and R. E. Kalman, Algebraic structure of linear dynamical systems, III: realization theory over a commutative ring, Proc. Nat. Acad. Sci. 69, 1972, 3404-3406.

17. R. W. Brockett and J. L. Willems, Systems defined on modules and applications to the control of linear partial differential equations, Automatica, to appear.

18. R. deB. Johnson, Linear systems over various rings, Report ESL-R-497, MIT, 1973.

19. B. F. Wyman, Linear systems over rings of operators, this volume.

20. R. W. Brockett and A. S. Willsky, Finite state homomorphic sequential machines, IEEE Trans. Auto. Cont. AC-17, 1973, 483-490.

21. B. D. O. Anderson, M. A. Arbib and E. G. Manes, Finitary and infinitary properties in categorical realization theory, to appear.

22. L. Schwartz, Théorie des Distributions, 2 vols., Hermann, 1950,51.

23. R. W. Newcomb, Linear Multiport Synthesis, McGraw-Hill, 1965.

24. R. E. Kalman and M. L. J. Hautus, Realization of continuous-time linear dynamical systems: rigorous theory in the style of Schwartz, Proc. NRL Conf. on Differential Equations, 1971.

25. E. W. Kamen, On an algebraic representation of continuous-time systems,

26. M. R. Wohlers, Lumped and Distributed Passive Networks, Academic Press, 1969.

(See also issues of the IEEE Transactions on Circuit Theory for material on lumped-distributed systems).

27. E. W. Kamen, Control of linear continuous-time systems defined over rings of distributions, this volume.

A CATEGORIST'S VIEW OF AUTOMATA AND SYSTEMS

M. A. Arbib
Computer and Information Science

E. G. Manes
Mathematics

University of Massachusetts
Amherst, MA 01002, U.S.A.

For $X : \mathcal{K} \to \mathcal{K}$ an endofunctor, the category $\mathrm{Dyn}(X)$ of <u>X-dynamics</u> has as objects pairs (Q,δ) with $\delta : QX \to Q$ and morphisms $f : (Q,\delta) \to (Q',\delta')$ --called <u>dynamorphisms</u>--defined by

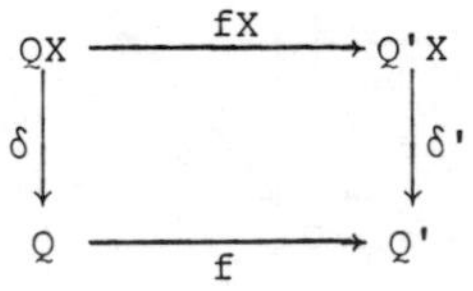

We say X is an <u>input process</u> if the forgetful functor $\mathrm{Dyn}(X) \to \mathcal{K}$ has a left adjoint (and hence is monadic [42], i.e., tripleable [15], [25], or algebraic [43], the Beck conditions being trivial). A <u>machine in</u> $\mathcal{K}$ ([7]) is $M = (X,Q,\delta,I,\tau,Y,\beta)$ where X is an input process, (Q,δ) is an X-dynamics, $\tau : I \to Q$ and $\beta : Q \to Y$ are morphisms in $\mathcal{K}$. Q is the <u>state object</u>, I is the <u>initial state object</u>, τ is the <u>initial state</u>, Y is the <u>output object</u> and β is the <u>output map</u>.

If $X : \mathcal{K} \to \mathcal{K}$ is an input process and A is an object of $\mathcal{K}$, let $(AX^{@}, A\mu_0, A\eta)$ be the free X-dynamics over A, i.e.,

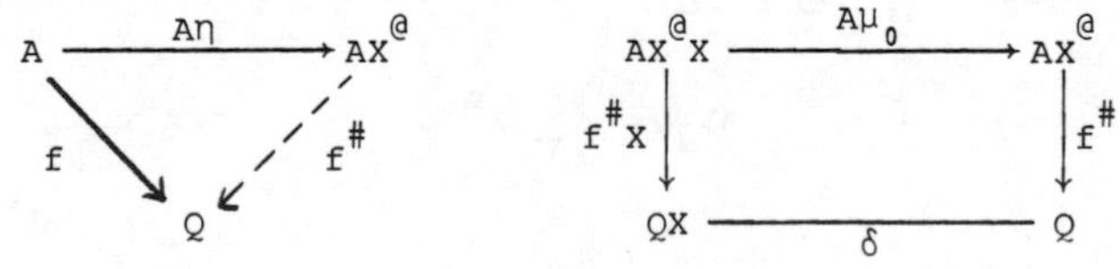

each X-dynamic-valued morphism f from A admits a unique dynamorphic extension $f^{\#}$. If M is a machine in $\mathcal{K}$, the reachability map $r : IX^@ \to Q$ and the run map $\delta^@ : QX^@ \to Q$ are defined as the dynamorphic extensions

$$\begin{array}{ccc} I & \xrightarrow{I\eta} & IX^@ \\ & {\scriptstyle\tau}\searrow \quad \swarrow{\scriptstyle\tau^{\#} = r} & \\ & Q & \end{array} \qquad\qquad \begin{array}{ccc} Q & \xrightarrow{Q\eta} & QX^@ \\ & {\scriptstyle id}\searrow \quad \swarrow{\scriptstyle id^{\#} = \delta^@} & \\ & Q & \end{array}$$

$IX^@$ is the object of inputs (see the example "sequential automata theory" below). The monad in $\mathcal{K}$ induced by the adjunction $Dyn(X) \rightleftarrows \mathcal{K}$ is $(X^@,\eta,\mu)$ where $\mu = (\mu_0)^@$. $X^@$ is the free monad over X via $\Gamma : X \to X^@$ defined by

$$\Gamma = X \xrightarrow{\eta X^@} X^@X \xrightarrow{\mu_0} X^@$$

It is shown in [14] that--so long as X and $\mathcal{K}$ satisfy reasonable rank conditions--the converse is true: if X generates a free monad, X is an input process. The diagram

$$\begin{array}{ccc} QX & \xrightarrow{Q\Gamma} & QX^@ \\ & {\scriptstyle\delta}\searrow \quad \swarrow{\scriptstyle\delta^@} & \\ & Q & \end{array}$$

establishes the expected bijective correspondence between X-dynamics and $(X^@,\eta,\mu)$-algebras ([14], [1]).

Let $\mathcal{K}$ be the category Set of sets and functions. Let Ω be an operator domain with rank at most the cardinal α, so that Ω is the disjoint family $(\Omega_n : n < \alpha)$ of sets (where Ω_n is the "set of n-ary operator labels of Ω"). An Ω-algebra (cf. [23], [36], [47] and [35]) is a pair (Q,δ) where Q is a set and δ assigns to each $\omega \in \Omega_n$ a function $\delta_\omega : Q^n \to Q$. An Ω-homomorphism $f : (Q,\delta) \to (Q',\delta')$ satisfies $(q_i : i < n)\delta_\omega f = (q_i f : i < n)\delta_\omega{}'$. Then the category of Ω-algebras is just $Dyn(X_\Omega)$ where X_Ω : Set $\to$ Set is defined as the coproduct

$$AX_\Omega = \coprod_{\omega\in\Omega_n} A^n$$

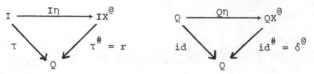

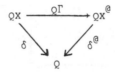

By the well known existence of free Ω-algebras (see [47], [43] for the infinitary case) X_Ω is an input process. No concrete example of an input process $X : \underline{Set} \to \underline{Set}$ is known which is not naturally equivalent to X_Ω for some Ω. We might mention the following counterexample which shows that any attempt to prove that every input process is X_Ω needs to be subtle: Let X be the functor which sends Q to the set of all ultrafilters on Q which are closed under countable intersections. Then X preserves countable coproducts and, so, is an input process with $QX^@$ the countable coproduct of all QX^n [8], [32]. The so-called "existence of a measurable cardinal" axiom may be phrased: "X is not naturally equivalent to the identity functor". If X is not the identity functor then X does not preserve all coproducts (so is not of the form $- \times X_0$) and hence is not of the form X_Ω (since X_Ω does not preserve even binary coproducts unless all Ω-operations are unary). On the other hand, it is not known at this writing whether there exists a model of Zermelo-Frankel set theory in which there exists a measurable cardinal. In short, a proof that every input process is X_Ω would settle a difficult open question of axiomatic set theory.

Sequential automata theory ([6]) is the case $\mathcal{K} = \underline{Set}$, $I = 1$ and $X = X_\Omega$ where $\Omega_1 = X_0$, $\Omega_n = \phi$ for $n \neq 1$. Thus $X = - \times X_0 : \underline{Set} \to \underline{Set}$. The interpretation of a dynamics $\delta : Q \times X_0 \to Q : q \to qx$ is that the system transfers to state qx if it was in state q when input x was applied. $\tau = q_0 \in Q$ is the initial state. $AX^@ = A \times X_0^*$ where X_0^* is the free monoid generated by X_0, and $\mu_0 : A \times X_0^* \times X_0 \to A \times X_0^*$ sends $x_1 \cdots x_n, x$ to $x_1 \cdots x_n x$. The reachability map $r : X_0^* \to Q$ sends $x_1 \cdots x_n$ to $q_0 x_1 \cdots x_n$, whereas the run map $\delta^@ : Q \times X_0^* \to Q$ sends $q, x_1 \cdots x_n$ to $q x_1 \cdots x_n$. The traditional inductive definition of, e.g., the run map is

(basis) $\quad (q,\Lambda)\delta^@ = q \quad$ (Λ is the empty word)

(inductive) $\quad (q, x_1 \cdots x_n x)\delta^@ = (q,x_1 \cdots x_n)\delta^@ x$

This tradition is preserved by the diagrams:

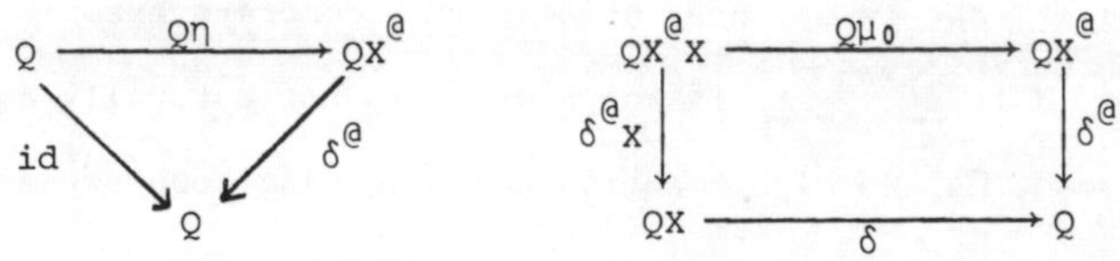

("basis") ("inductive")

Tree (or algebra) automata theory is the case $\mathcal{K}$ = Set, $X = X_\Omega$ for finitary Ω, that is, $\Omega_n = \phi$ if n is an infinite cardinal. For further discussion see [7], section 5 of the first part of this introduction and the paragraph on "syntax-directed translation" below.

(Discrete, time-invariant) linear control (or systems) theory [37] is the case $\mathcal{K}$ = Vect, the category of real vector spaces and linear maps, and $X = id$: Vect $\to$ Vect. We pause to explain why (see also part 2 of this introduction). Consider the physical system

$$\dot{q} = f(q,x)$$

where $f : \mathbb{R}^n \times \mathbb{R}^m \to \mathbb{R}^n$, $q \in \mathbb{R}^n$ is the "state vector" of the system and $x \in \mathbb{R}^m$ is the "input vector" of (environmental and control) forces acting on the system. q might be a 6-vector representing the position and velocity of a particle in 3-space. For "textbook" examples refer to [45, 2.5]. The system is explicitly time-invariant, i.e., we have not written $\dot{q} = f(q,x,t)$; but the time-varying case is in fact still an example of the general theory, as discussed in [10]. We make the

Discretization assumption: There is a fixed sampling interval of time, Δt, such that $\dot{q}(t) = (q(t+\Delta t) - q(t))/\Delta t$ is true for "practical purposes".

The discretization assumption is particularly viable for modern control systems in which input and state are monitored by digital computer. Writing all times as integral multiples of the "fundamental quantum" $\Delta t = 1$, we have $q(t+1) = q(t) + \dot{q}(t)$. Expanding f in a Taylor series about $(0,0)$ ([45, 3.4]) yields $f(q,x) = c + Aq + Bx + qCx$ + higher order terms. Coordinates may be chosen to force $c = 0$. In much (but not all!) of control theory, it has proven useful to make the

Linearity assumption: The approximation $f(q,x) = Aq + Bx$ is true for "practical purposes".

Putting it all together, we have

$$q(t+1) = (A + id)q + Bx$$

that is, a dynamics (with respect to $id : \mathbf{Vect} \to \mathbf{Vect}$) $\delta = A + id : \mathbb{R}^n \to \mathbb{R}^n$, and an initial state $\tau = B : \mathbb{R}^m \to \mathbb{R}^n$. "Reachability" and other general machine concepts (such as "observability") as discussed below are important in qualitative linear analysis as taught to undergraduates, see, e.g., [45]. For more perspective, see part 2 of this introduction.

Let $M = (X,Q,\delta,I,\tau,Y,\beta)$ be a machine in $\mathcal{K}$. The response or behavior of M (as seen externally) is $f_M = r\beta : IX^@ \to Y$. The converse problem (for fixed I, X, Y), given arbitrary $f : IX^@ \to Y$, is to "build" M such that M realizes f (i.e., $f_M = f$) and M is "optimal" with this property in the sense that "all states are used" and "no two states have the same effect". Let $(\mathcal{E},\mathcal{M})$ be an image factorization system for $\mathcal{K}$ (introduction, 3.3). The machine M "uses all states" if M is reachable, i.e., $r : IX^@ \to Q$ is in $\mathcal{E}$. A simulation $\psi : M \to M'$ is a dynamorphism $\psi : (Q,\delta) \to (Q',\delta')$ which commutes with input and output:

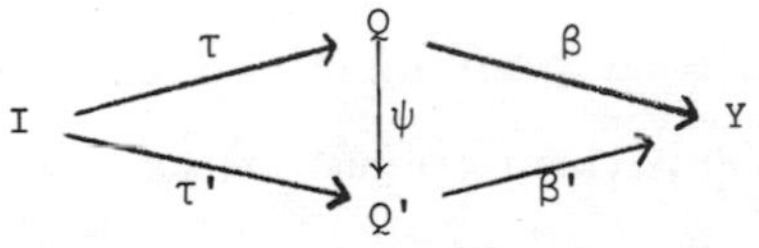

Notice that the existence of a simulation $M \to M'$ implies that $f_M = f_{M'}$. A minimal realization of f is a terminal object $M_f = (X,Q_f,\delta_f,I,\tau_f,Y,\beta_f)$ in the category of reachable realizations of f and simulations. Thus a reachable realization M admits a simulation $\psi : M \to M_f$ (such ψ being necessarily in $\mathcal{E}$) which "merges states with the same effect". One way to construct M_f in the case of sequential machine theory is by "Nerode equivalence" (see [6]). A categorical approach ([7], [5]) is as follows.

Given $f : IX^@ \to Y$, the <u>Nerode equivalence of</u> f, if it exists, is $(E_f, \alpha_f, \gamma_f)$ where $\alpha_f, \gamma_f : E_f \to IX^@$ satisfies $(\alpha_f)^{\#}f = (\gamma_f)^{\#}f : E_fX^@ \to Y$ and is universal with this property:

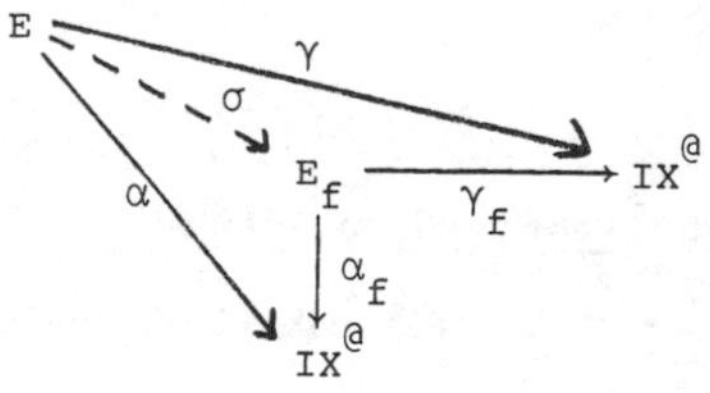

If $\alpha^{\#}f = \gamma^{\#}f$ there exists unique σ as shown.

Under favorable conditions (here $\mathcal{E}$ = coequalizers), all of which hold for the contexts of sequential, tree and linear automata theory, $r_f : IX^@ \to Q_f = \mathrm{coeq}(\alpha_f, \gamma_f)$ is built up to the reachability map of M_f. In the tree automata case, this is a new result for $I \neq \phi$ ([5]). In all three cases, E_f is a kernel pair. However: if $X = X_\Omega$ where Ω has at least one infinitary operation, $I = 1$, $Y = 2$ and $f : IX^@ \to Y$ is the characteristic function of the set of "trees with a finite derivation," then f has no minimal realization ([5]). This suggests that the existence of minimal realizations--as yet to be regarded as an unsolved problem of category theory--is a "finiteness" property (characteristic to the machine point of view) and is not just a consequence of completeness and cocompleteness.

An interesting situation in categorical machine theory--which includes sequential and linear automata, but excludes tree automata--occurs when $\mathcal{K}$ has countable products and coproducts and $X : \mathcal{K} \to \mathcal{K}$ has a right adjoint $X^\cdot : \mathcal{K} \to \mathcal{K}$ ([8]; see also [32], [12], [26]). Call this <u>adjoint</u> machine theory. In this case, $\mathrm{Dyn}(X) \to \mathcal{K}$ has both a left adjoint $A \mapsto (AX^@, A\mu_0)$ and a right adjoint $A \mapsto (AX_@, AL)$ where $AX^@$ is the classical free monoid construction, $AX^@ = \coprod_{n=0}^{\infty} AX^n$ and, dually, $AX_@ = \prod_{n=0}^{\infty} A(X^\cdot)^n$. Let $M = (X, Q, \delta, I, \tau, Y, \beta)$ be a machine in $\mathcal{K}$.

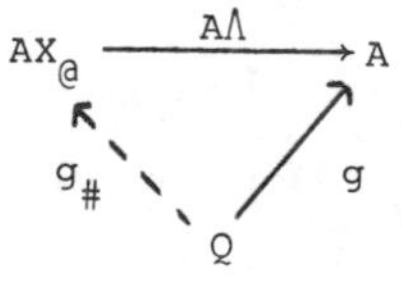

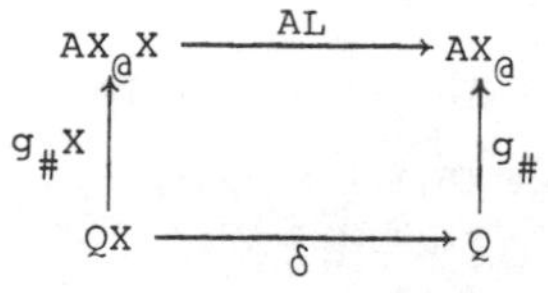

the universal property of the cofree dynamics

In the sequential machine case, $YX^{@}$ is the set of functions from X_0^* to Y, i.e., the "response space" (when $I = 1$); $Y\Lambda : YX^{@} \to Y$ evaluates at the empty word, and the dynamical structure $YL : YX^{@} \times X_0 \to YX^{@}$ is the "left shift" $f,x \to L_x f$ where $wL_x = xw \in X_0^*$. In general, the <u>observability map of</u> M is defined to be the dynamorphic coextension of the output map:

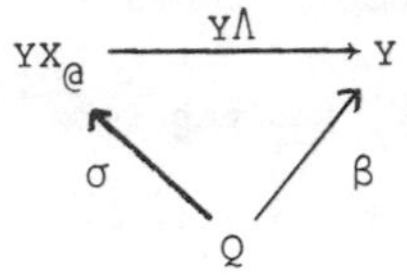

(basis)

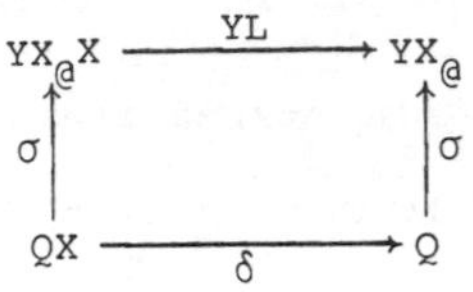

(inductive)

We leave to the reader the exercise of discovering the traditional inductive definition of σ for sequential machines; here, $q\sigma$ is "the response of M if τ were q" and represents what can be learned about q by experiment. For interpretations of $X_{@}$ and observability see [8] and [37].

In the context of adjoint machines, a realization of f is minimal if and only if it is reachable and observable (M is <u>observable</u> if $\sigma \in \mathcal{M}$, that is "different states can be distinguished by experimentation"). So long as $X^{@}$ preserves $\mathcal{E}$ (which is likely since $X^{@}$ has $X_{@}$ as a right adjoint) M_f exists; Q_f may be constructed as the $\mathcal{E}$-$\mathcal{M}$ factorization of the dynamorphic coextension $(f_M)_\# : IX^{@} \to YX_{@}$ of $f_M : IX^{@} \to Y$.

The adjointness relation

$$\frac{QX \xrightarrow{\ \delta\ } Q}{Q \xrightarrow{\ \delta^{\cdot}\ } QX^{\cdot}}$$

allows us to define the machine $M^{op} = (X^{\cdot}, Q, \delta^{\cdot}, Y, \beta, I, \tau)$ in $\mathcal{K}^{op}$. We have $(M^{op})^{op} = M$. We also have [8]

<u>Fundamental duality metatheorem for adjoint machines</u>.

The reachability map of M is the observability map of M^{op}.

The traditional duality situation ([37], cf. [48]) is more highly structured. While most examples of adjoint machines occur when $\mathcal{K}$ is a closed category **(this introduction, part 1)** and $X = - \otimes X_0$ --directly mimicking sequential machine theory--an attempt to formalize this (as have [32], [26]) sacrifices the duality theory since the dual of a closed category need not be closed.

We remark that much of the abstract theory of adjoint machines, including minimal realization and duality, goes through for a functor $\mathcal{L} \to \mathcal{K}$ with two adjoints, whether or not $\mathcal{L}$ has the form Dyn(X). This is in fact implicit in the work of Bainbridge [12], [13]. It is not easy to imagine a machine-like example that is not monadic, but why need the monad be free? Indeed, a non-free example is provided by the S/k-systems of Wyman [55]. **For yet another example, see [9].**

Syntax-directed translation has been studied in mathematical linguistics (transformational theory) ([22]) and in programming semantics ([28]). A formalization using **Appelgate** triple maps ([4]) has been developed in [3]. Specializations include tree automata, "generalized sequential machines" ([29]) and "generalized^2 sequential machines" [50]. Combined with the theory of non-deterministic machines [9] (see also [20], [30], [33] and [27]), a proper formulation of stochastic automata ([46]) can be achieved. There are both direct- and inverse-state translations. The following is inverse-state:

Example: a program to differentiate $x + \sin(x)$

Linguistic formulation:

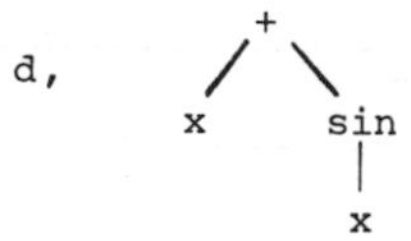

rules:

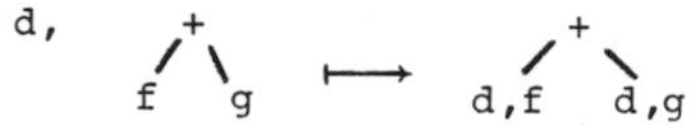

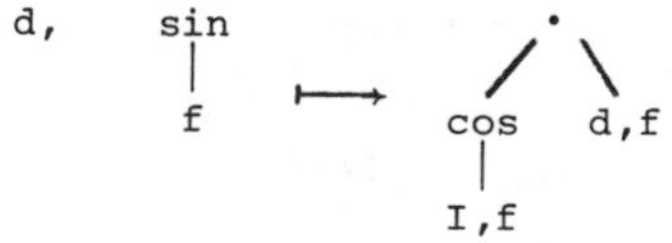

(I = do nothing)

$$
\begin{array}{ccc}
\text{I,} & \begin{array}{c} + \\ \diagup\ \diagdown \\ f \quad g \end{array} & \longmapsto \quad \begin{array}{c} + \\ \diagup\ \diagdown \\ \text{I,}f \quad \text{I,}g \end{array}
\end{array}
$$

$$
\begin{array}{ccc}
\text{I,} & \begin{array}{c} \sin \\ | \\ f \end{array} & \longmapsto \quad \begin{array}{c} \sin \\ | \\ \text{I,}f \end{array}
\end{array}
$$

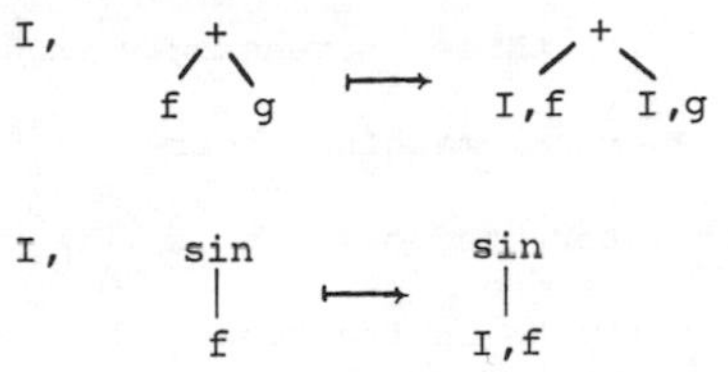

Thatcher-Style ([50]) formulation: Define the set $Q = \{d, I\}$ of "states" and the "translation"

$$
Q \times ((A^2 \times \{+\}) + (A \times \{\sin\})) \xrightarrow{\;AT\;} (Q \times A)T
$$

where T is the monad induced by binary $+, \cdot$ and unary $\sin$.

Alagić formulation ([1], [3]): A translation is defined to be a natural transformation $\tau : QX \to TQ$ where X is an input process, T is a monad and Q is a "state functor" which has a right adjoint (i.e., the set Q has become $Q \times -$). Among other results, there exists a canonical extension $\bar{\tau} : QX^{@} \to TQ$ which "runs" the translation.

A plethora of other categorical ideas have been introduced in the system sciences. While the concept of "algebra over a monad" is an adequate generalization of the concept of a monoid acting on a set for many purposes, there is another point of view motivated by re-viewing an M-set as a functor from M (qua one-object category) to Set: automata are Set-valued functors. This point of view is quite useful for "state graphs" and "system interconnection", and ties in well with algebras as originally defined by Lawvere ([40]) (but which are still coextensive with algebras over a monad). Cf. [12], [16], [21], [26], [31], [34].

Recent interest in cartesian closed categories (section 6 of part 1 of this introduction) has spilled over into automata theory. Lambek's interpretation of free cartesian closed categories as deductive calculi [39] led to a "Gentzen cut theorem" which inspired Wand [51], [52]. The fact that Scott's category of continuous lattices [49] is cartesian closed has led to axiomatizations in the theory of programming languages that emphasize this point of view ([2], [53], [54]).

We close this introduction with three increasingly vague speculations upon the immediate future of categorical machine theory.

In actual contexts, one is interested only in $f : IX^@ \to Y$ which are "finite-state" [6] (e.g., Q_f is finite in the sequential case and finite-dimensional in the linear case). A general theory of finite-state systems is fraught with ambiguity . . . there are many possible approaches. The first such paper is [5]. The problem is well-formulated and we look forward to a distillation of the basic principles (hopefully the work of many specialties) in coming years.

Non-linear systems pose a problem which is universally regarded as essentially unsolved. We optimistically hope that the categorical approach will provide insight into the key concepts which are apparently missing. In this connection we mention an intriguing example of Goguen's ([32]) which has not been adequately analyzed. Let $\mathcal{K}$ be the category of affine vector spaces ($f - f(0)$ is linear). $\mathcal{K}$ is a closed category with $X \otimes_A Y = X \otimes Y + X + Y$. The dynamics $\delta : Q \otimes_A X_0 \to Q$ corresponds to the less truncated Taylor expansion

$$\dot{q} = c + Aq + Bx + qCx$$

The object of inputs $IX^@ = I + (I \otimes_A X_0) + (I \otimes_A X_0 \otimes_A X_0) + \cdots$ is quite complicated and not likely to arise on the basis of engineering intuition which has always favored "$IX^@ = X_0^*$", i.e., "the most general input is a sequence of primitive inputs". (In the context of the group machines of [17], [18] the categorical approach again leads the choice for the object of inputs away from the weak direct sum group structure on X_0^* --which produces "reachability" maps which are not group homomorphisms-- to the countable copower $X_0^§$ which is a very different sort of group). A fascinating open question is: for what sorts of truncated Taylor series is there a category $\mathcal{K}$ whose objects include finite-dimensional vector spaces, and an input process X in $\mathcal{K}$ for which X-machines model systems $\dot{q} = f(q,x)$ with f truncated in the specified way?

As a final commentary, we seek the proper role of the Goguen-Ehrig-Pfender closed category machines. To begin, let us observe that any machine $M = (X,Q,\delta,I,\tau,Y,\beta)$ in $\mathcal{K}$ induces the sequential machine with $X_0 = (I,IX)$, states (I,Q), dynamics

$$(I,Q) \times X_0 \longrightarrow (I,Q)$$

$$q \ , \ x \ \mapsto \ I \xrightarrow{x} IX \xrightarrow{qX} QX \xrightarrow{\delta} Q$$

initial state τ and output $- \circ \beta : (I,Q) \to (I,Y)$. This passage describes a functor from Dyn(X) to Dyn($- \times X_0$) which is limit preserving and so, consequently, is often expected to have a left adjoint (by the adjoint functor theorem, see, e.g., [42], [11]). There is also a similar functor from Dyn(X) to M-sets, obtained by replacing X with $X^@$; here, $M = \mathcal{K}(I,IX^@) \cong \text{Dyn}(X)(IX^@,IX^@)$ is a monoid under composition. In short, all machines admit a <u>classical comparison</u> further study of which is yet to be done. Recent work in closed category theory [24] and closed monad theory [38], [41], where a closed category $\mathcal{V}$ is the universe of discourse (Dubuc speaks of "the $\mathcal{V}$-world"), has emphasized Kan extension techniques which relativize all ordinary concepts (the "<u>Set</u>-world") wuch as adjointness ("$\mathcal{V}$-adjointness"). We submit, then, that the future will reveal $- \otimes X_0$ machines in the closed category $\mathcal{V}$ to be the "classical comparisons" of $\mathcal{V}$-automata in $\mathcal{V}$-categories (a $\mathcal{V}$-<u>category</u> is a category whose hom objects are not sets, but are objects in $\mathcal{V}$). Automata in additive categories have been considered (cf. [19], [44]); but additive categories are just $\mathcal{V}$-categories where $\mathcal{V}$ is the closed category of abelian groups.

REFERENCES

1. S. Alagić, Natural state transformations, Department of Computer and Information Science Technical Report 73B-2, Univ. of Mass., 1973.

2. --- Algebraic aspects of ALGOL 68, Computer and Information Science Technical Report 73B-5, Univ. of Mass., 1973.

3. --- Categorical theory of tree transformations, this volume.

4. H. Appelgate, Acyclic models and resolvent functors, dissertation, Columbia University, 1965.

5. B. D. O. Anderson, M. A. Arbib and E. G. Manes, Finitary and infinitary conditions in categorical realization theory, to appear.

6. M. A. Arbib, Theories of Abstract Automata, Prentice-Hall, 1969.

7. M. A. Arbib and E. G. Manes, Machines in a category: an expository introduction, Siam Review, to appear.

8. --- Adjoint machines, state-behavior machines, and duality, J. Pure Appl. Alg., to appear.

9. --- Fuzzy morphisms in automata theory, this volume.

10. --- Time-varying systems, this volume.

11. --- The Categorical Imperative: Arrows, Structures, and Functors, Academic Press, 1974.

12. E. S. Bainbridge, A unified minimal realization theory, with duality for machines in a hyperdoctrine, dissertation, Univ. of Mich., 1972.

13. --- Addressed machines and duality, this volume.

14. M. Barr, Coequalizers and free triples, Math. Zeit. 116, 1970, 307-322.

15. J. Beck, Triples, algebras and cohomology, dissertation, Columbia University, 1967; available from University Microfilms, Ann Arbor, Michigan.

16. D. Benson, An abstract machine theory for formal language parsers, this volume.

17. R. W. Brockett and A. S. Willsky, Finite-state homomorphic sequential machines, IEEE Trans. Aut. Cont. AC-17, 1973, 483-490.

18. --- Some structural properties of automata defined on groups, this volume.

19. L. Budach, Automata in additive categories with applications to stochastic linear automata, this volume.

20. E. Burroni, Algèbres relatives à une loi distributive, Comp. R. Acad. Sc. Paris, t. 276 (5 fév. 1973), Ser. A, 443-446.

21. R. W. Burstall and J. W. Thatcher, Algebraic theory of recursive program schemes, this volume.

22. N. Chomsky, On certain formal properties of grammars, Inf. Cont. 2, 1959, 136-67.

23. P. M. Cohn, Universal Algebra, Harper and Row, 1965.

24. E. J. Dubuc, Kan Extensions in Enriched Category Theory, Lecture Notes in Mathematics 145, Springer-Verlag, 1970.

25. B. Eckmann ed., Seminar on Triples and Categorical Homology Theory, Lecture Notes in Mathematics 80, Springer-Verlag, 1969.

26. H. Ehrig; K. D. Kiermeier; H.-J. Kreowski and W. Kühnel, Systematisierung der Automatentheorie, Seminarbericht, Technische Universität Berlin, Fachbereich Kybernetik, 1973.

27. H. Ehrig and H.-J. Kreowski, Power and initial automata in pseudoclosed categories, this volume.

28. E. Engeler ed., Symposium on Semantics of Algorithmic Languages, Lecture Notes in Mathematics 188, Springer-Verlag, 1971.

29. S. Ginsburg, An Introduction to Mathematical Machine Theory, Addison-Wesley, 1962.

30. J. A. Goguen, L-fuzzy sets, J. Math. Anal. Appl. 18, 1967, 145-174.

31. --- On homomorphisms, simulations, correctness, subroutines, and termination for programs and program schemes, Proceedings of the 13th IEEE Conference on Switching and Automata Theory, 1972, 52-60.

32. --- Minimal realization of machines in closed categories, Bull. Amer. Math. Soc. 78, 1972, 777-783.

33. --- Axioms, extensions and applications for fuzzy sets: languages and the representation of concepts, to appear.

34. --- Some comments on applying mathematical system theory, to appear.

35. J. A. Goguen; J. W. Thatcher; E. G. Wagner and J. B. Wright, A junction between computer science and category theory, I: basic concepts and examples (part 1), IBM Research Report RC 4526, T. J. Watson Research Center, 1973.

36. G. Grätzer, Universal Algebra, Van Nostrand, 1967.

37. R. E. Kalman; P. L. Falb and M. A. Arbib, Topics in Mathematical System Theory, McGraw-Hill, 1969.

38. A. Kock, Monads on symmetric monoidal closed categories, Arch. Math. XXI, 1970, 1-10.

39. J. Lambek, Deductive systems and categories, Math. Sys. Th. 2, 1968, 287-318.

40. F. W. Lawvere, Functorial semantics of algebraic theories, dissertation, Columbia University, 1963.

41. F. E. J. Linton, Triples for closed categories, Seminar in Category Theory, notes by D. Taylor, NSF summer institute at Bowdoin College, 1969.

42. S. Mac Lane, Categories for the Working Mathematician, Springer-Verlag, 1972.

43. E. G. Manes, Algebraic Theories, Springer-Verlag, to appear.

44. J. Meseguer and I. Sols, Automata in semimodule categories, this volume.

45. L. Padulo and M. A. Arbib, System Theory: a Unified Approach to Discrete and Continuous Systems, W. B. Saunders, 1974.

46. A. Paz, Introduction to Probabilistic Automata, Academic Press, 1971.

47. R. S. Pierce, Introduction to the Theory of Abstract Algebras, Holt, Rinehart and Winston, 1968.

48. J. Rissanen and B. F. Wyman, Duality of input/output maps, this volume.

49. D. Scott, The lattice of flow diagrams, Symposium on Semantics of Algorithmic Languages, Lecture Notes in Mathematics 188, Springer-Verlag, 1971.

50. J. W. Thatcher, Generalized^2 sequential machine maps, J. Comp. Syst. Sci. 4, 1970, 339-367.

51. M. Wand, An unusual application of program-proving, Proceedings of the Fifth ACM Symposium on the Theory of Computing, 1973, 59-66.

52. --- An algebraic formulation of the Chomsky hierarchy, this volume.

53. --- On the recursive specification of data types, this volume.

54. --- On the behavioral description of data structures, to appear.

55. B. F. Wyman, Linear systems over rings of operators, this volume.

CATEGORICAL THEORY OF TREE PROCESSING[†]

Suad Alagić
Computer and Information Science
University of Massachusetts
Amherst, Mass. 01002

The basic algebraic structure in the theory of sequential machines is $X_0{}^*$, the free monoid on the set X_0 of generators. In this paper we show that the corresponding structure for the theory of tree processing is not an ordinary monoid, but a monad [15, Chap. VI]. Introducing monads we generalize the theory of tree transformations as presented in [10], [22], etc., in such a way that the theory also includes generalized sequential machines (cf. [12], p. 93) as a particular case. For more details see [1].

Let Σ be a *ranked alphabet* (cf. [13], [21]). Σ can be interpreted as a functor $\underline{\mathrm{Set}} \to \underline{\mathrm{Set}}$ where ΣA is the set of all formal expressions of the form $\sigma[<a_1>,\ldots,<a_n>]$ (σ is an n-ary symbol) and for $f : A \to A'$ $\Sigma f(\sigma[<a_1>,\ldots,<a_n>]) = \sigma[<f(a_1)>,\ldots,<f(a_n)>]$. A *$\Sigma$-algebra* can be defined as a pair (A,δ) where $\delta : \Sigma A \to A$ is a function. A *homomorphism* $f : (A,\delta) \to (A',\delta')$ *of Σ-algebras* is a function $f : A \to A'$ such that $f \cdot \delta = \delta' \cdot \Sigma f$. Denote with $\underline{\Sigma\text{-alg}}$ the category of Σ-algebras. $\underline{\Sigma\text{-alg}}$ is a particular case of the category $\underline{\mathrm{Dyn}(X)}$ (cf. [2]). $X : \underline{C} \to \underline{C}$ is a functor,

[†]The research reported in this paper was supported in part by the National Science Foundation under Grant No. GJ 35759. The author gratefully acknowledges the help which he received from Dr. Michael Arbib and Dr. Ernest Manes and the comments given by Dr. Brian Anderson and Mr. Steven Hegner.

and Dyn(X) has as objects pairs (Q,δ) where Q is an object of the category $\underline{C}$ and $\delta : XQ \to Q$ is a morphism of $\underline{C}$. A morphism $f : (Q,\delta) \to (Q,\delta')$ in Dyn(X) is a morphism $f : Q \to Q'$ of $\underline{C}$ such that $f \cdot \delta = \delta' \cdot Xf$. Dyn($\Sigma$) is precisely Σ-alg. A special case is Dyn($-\times X_0$) with objects (Q,δ) where $\delta : Q \times X_0 \to Q$.

It is a well known fact that given a set Z we can construct a *free Σ-algebra* $(T_{\Sigma,Z}, \mu_0 Z)$ *on* Z *generators* (cf. [13] and [22]). Here $T_{\Sigma,Z}$ is the set of *Σ-trees on* Z *generators* (cf. [22]). Notice that ΣZ as defined before is the set of one level Σ-trees on Z generators. The construction of a free Σ-algebra is in fact a functor $F : \underline{\text{Set}} \to \underline{\Sigma\text{-alg}}$ which sends every set Z to $(T_{\Sigma,Z}, \mu_0 Z)$. F has a right adjoint, the forgetful functor $U : \underline{\Sigma\text{-alg}} \to \underline{\text{Set}}$ sending a Σ-algebra (A,δ) to its underlying set A.

Recall that an *adjunction* $(F,U,\eta,\varepsilon) : \underline{A} \to \underline{B}$ consists of a pair of functors F,U where $F : \underline{A} \to \underline{B}$ and $U : \underline{B} \to \underline{A}$, and a pair of natural transformations $\eta : I_{\underline{A}} \to UF$, $\varepsilon : FU \to I_{\underline{B}}$ such that $U\varepsilon \cdot \eta U = 1_U$ and $\varepsilon F \cdot F\eta = 1_F$ (cf. [15], pp. 78-81). I denotes the identity functor. F is said to be a left adjoint to U and U a right adjoint to F.

In general, when $U : \text{Dyn}(X) \to \underline{C}$ has a left adjoint $F : \underline{C} \to \text{Dyn}(X)$ we call X an *input process*. Both Σ and $-\times X_0$ are input processes. If X is an input process we denote $FQ = (X^@Q, \mu_0 Q)$ where $X^@ = UF$. $(X^@Q, \mu_0 Q)$ is called a *free dynamics* (cf. [2]) *on* Q *generators*. Examples are $(Q \times X_0{}^*, \mu_0 Q)$ where $\mu_0 Q : (q,w,x) \to (q,wx)$ for ordinary sequential machines and $(T_{\Sigma,Q}, \mu_0 Q)$ in the tree case.

The existence of an adjunction gives rise to a triple (T,η,μ) where $T = UF$ is a functor and $\mu = U\varepsilon F : TT \dot{\to} T$ and $\eta : I \dot{\to} T$ are natural transformations. (T,η,μ) has a monoid-like structure and is called a *monad* [15, Chap. VI]. It satisfies $\mu \cdot \mu T = \mu \cdot T\mu$ (the associativity axiom), $\mu \cdot \eta T = 1_T$ and $\mu \cdot T\eta = 1_T$ (the unitary axioms). A *morphism* $(T,\eta,\mu) \to (\bar{T},\bar{\eta},\bar{\mu})$ *of monads* is a natural transformation $\bar{\tau} : T \dot{\to} \bar{T}$ such that $\mu \cdot \bar{\tau} = \bar{\mu} \cdot \bar{\tau}\bar{T} \cdot T\bar{\tau}$ and $\bar{\tau} \cdot \eta = \bar{\eta}$.

The adjunction $\underline{Set} \underset{U}{\overset{F}{\rightleftarrows}} \underline{\Sigma\text{-alg}}$ constructs the *tree monad* (T,η,μ) where $T : \underline{Set} \to \underline{Set}$ is the *tree functor* defined on objects as $TZ = T_{\Sigma,Z}$. η is defined on elements as $\eta Z : z \to \langle z\rangle$, i.e., it sends a variable z to the one node tree $\langle z\rangle$, and μ removes one (outer) level of brackets, so that it maps the set TTZ of Σ-trees on Σ-trees as generators to the set TZ of Σ-trees on Z-generators. For $f : Z \to Y$, Tf is defined to be the substitution of generators according to f. Similarly, for the input process $-\times X_0$ we get a monad determined by $T = -\times X_0{}^*$, $\mu : -\times X_0{}^*\times X_0{}^* \dot{\to} -\times X_0{}^*$ defined as concatenation and $\eta : I \dot{\to} -\times X_0{}^*$ defined on elements as $- \to (-,\Lambda)$ where Λ is the empty word.

If (A,δ) is a Σ-algebra then $\delta : \Sigma A \to A$ can be extended uniquely by algebraic recursion to $h : TA \to A$ where T is the tree functor. h satisfies $h{\cdot}Th = h{\cdot}\mu A$ and $h{\cdot}\eta A = 1_A$ and the pair (A,h) is called a T-algebra. Another example of a T-algebra is (Q,δ^*) where $\delta^* : Q\times X_0{}^* \to Q$. For a given monad (T,η,μ) in $\underline{C}$, let $\underline{C}^T$ denote the category of T-algebras. A morphism $f : (A,h) \to (A',h')$ of $\underline{C}^T$ is a morphism $f : A \to A'$ of $\underline{C}$ such that $f{\cdot}h = h'{\cdot}Tf$.

Using the language developed so far we can introduce the first of the two general models which in particular cases reduces to generalized sequential machines, deterministic bottom-up tree transformations and tree automata. We define a *direct state transformation on a free monad* as a natural transformation $\tau : XQ \dot{\to} Q\bar{T}$ where $X : \underline{C} \to \underline{C}$ is an input process with the associated *input monad* $(X^@,\eta,\mu)$, $(\bar{T},\bar{\eta},\bar{\mu})$ is a monad in $\underline{C}$ called the *output monad* and $Q : \underline{C} \to \underline{C}$ a functor called the *state functor*. Observing that a functor $X : \underline{C} \to \underline{C}$ determines a functor $X_\bullet : \underline{C}^{\underline{C}} \to \underline{C}^{\underline{C}}$ (objects of $\underline{C}^{\underline{C}}$ are functors $\underline{C} \to \underline{C}$ and morphisms natural transformations) defined to be just the composition with X, we can prove:

<u>Theorem</u> If X is an input process in $\underline{C}$ then $X_\bullet$ is an input process in $\underline{C}^{\underline{C}}$.

The above theorem implies:

Theorem The formulas $\bar{\tau}\cdot\eta Q = Q\bar{\eta}$ and $\bar{\tau}\cdot\mu_0 Q = Q\bar{\mu}\cdot\tau\bar{T}\cdot X\bar{\tau}$ define a unique extension of $\tau : XQ \dot{\to} Q\bar{T}$ to the natural transformation (the *state transformation*) $\bar{\tau} : X^{@}Q \dot{\to} Q\bar{T}$.

A generalized sequential machine $Y_0^* \xleftarrow{\lambda} Q\times X_0 \xrightarrow{\delta} Q$ becomes a direct state transformation $\tau : -\times Q\times X_0 \dot{\to} -\times Y_0^*\times Q$ which can be extended in a unique fashion to $\bar{\tau} : -\times Q\times X_0^* \dot{\to} -\times Y_0^*\times Q$. Another particular case is Engelfriet's bottom-up tree transformations (cf. [10]). For Σ and Ω ranked alphabets (and also interpreted as functors) and T and $\bar{T}$ the corresponding tree functors $(TZ = T_{\Sigma,Z},\ \bar{T}Z = T_{\Omega,Z})$ a bottom-up tree transformation is specified by a natural transformation $\tau Z : \Sigma(Z\times Q) \dot{\to} (\bar{T}Z)\times Q$ which can be extended uniquely to $\bar{\tau}Z : T(Z\times Q) \dot{\to} (\bar{T}Z)\times Q$. In particular, for $Z = \phi$ we get $\bar{\tau}\phi : T_{\Sigma} \to T_{\Omega}\times Q$. Tree automata are obtained as a particular case (cf. [10]).

The name of the transformation is justified since $(X^{@},\eta,\mu)$ is indeed a free monad. But the input monad does not have to be free. For arbitrary monads (T,η,μ) and $(\bar{T},\bar{\eta},\bar{\mu})$ in $\underline{C}$ and a functor $Q : \underline{C} \to \underline{C}$ define a *direct state transformation* as a natural transformation $\bar{\tau} : TQ \dot{\to} Q\bar{T}$ such that $\bar{\tau}\cdot\eta Q = Q\bar{\eta}$ and $\bar{\tau}\cdot\mu Q = Q\bar{\mu}\cdot\bar{\tau}\bar{T}\cdot T\bar{\tau}$. A direct state transformation $\bar{\tau} : TQ \dot{\to} Q\bar{T}$ is called *pure* when Q is the identity functor I. It follows from the definition that it is then just a morphism of monads.

Appealing to the comparison theorem (cf. [15], p. 138) we can prove:

Theorem The direct state transformation on a free monad $\bar{\tau} : X^{@}Q \dot{\to} Q\bar{T}$ is a direct state transformation.

In particular, the above theorem gives a result about bottom-up tree transformations which is dual to Thatcher's (cf. [22], p. 353). The theorem also implies:

Corollary $(X^{@},\eta,\mu)$ is a free monad (cf. [7]) on X generators with the natural transformation $\zeta = \mu_0\cdot X\eta$ as the inclusion of generators.

This in particular means that both the tree monad and the monad associated with $-\times X_0^*$ are free.

To recapture a monad as a monoid of actions define an *action of a monad* (T,η,μ) on the functor $Q : \underline{C} \to \underline{C}$ as a natural transformation $\nu : TQ \to Q$ such that $\nu \cdot \mu Q = \nu \cdot T\nu$ and $\nu \cdot \eta Q = 1_Q$. Then we have:

Proposition A direct state transformation $\bar{\tau} : (T,\eta,\mu) \to (I,1_I,1_I)$ is precisely an action of the monad (T,η,μ) on the state functor Q.

Let $\bar{\tau} : TQ \dot{\to} QK$ and $\bar{\rho} : KS \to SL$ be direct state transformations. Define the composite transformation $\bar{\rho} * \bar{\tau}$ as $Q\bar{\rho} \cdot \bar{\tau}S$.

Theorem The composition of direct state transformations is a direct state transformation.

Now we can introduce a model dual to direct state transformations. A particular case of the model are Thatcher's generalized[2] sequential machine maps (top-down tree transformations) (cf. [22]). We show that the nature of the duality of our two general models, bottom-up and top-down tree transformations being their particular cases, is an adjunction.

An *inverse state transformation on a free monad* is a natural transformation $\tau : QX \dot{\to} \bar{T}Q$ where X is an input process in $\underline{C}$ with the associated *input monad* $(X^@,\eta,\mu)$, $(\bar{T},\bar{\eta},\bar{\mu})$ is the *output monad* and $Q : \underline{C} \to \underline{C}$ the *state functor* with the property that it has a right adjoint $Q^{\bullet}$.

Theorem The formulas $\bar{\tau} \cdot Q\eta = \bar{\eta}Q$ and $\bar{\tau} \cdot Q\mu_0 = \bar{\mu}Q \cdot \bar{T}\bar{\tau} \cdot \tau X^@$ define a unique extension of $\tau : QX \dot{\to} \bar{T}Q$ to the natural transformation $\bar{\tau} : QX^@ \dot{\to} \bar{T}Q$ called the *state transformation*.

A part of the proof of the above theorem is the following proposition:

Proposition To every inverse state transformation $\bar{\tau} : QX^@ \dot{\to} \bar{T}Q$ corresponds a unique pure direct state transformation $\bar{\rho} : X^@ \dot{\to} L$ where $L = Q^{\bullet}\bar{T}Q$.

For Σ and Ω ranked alphabets (and interpreted as functors) and T and $\bar{T}$ the corresponding tree functors ($TZ = T_{\Sigma,Z}$ and $\bar{T}Z = T_{\Omega,Z}$) a top-down tree transformation is specified as a natural transformation $\tau Z : (\Sigma Z) \times Q \dot{\to} \bar{T}(Z \times Q)$. Observe that $-\times Q$ has a right adjoint $(-)^Q$.

According to the last theorem τZ can be extended uniquely to $\bar{\tau}Z : (TZ)\times Q \dot{\rightarrow} \bar{T}(Z\times Q)$ (compare to Thatcher [22], p. 352). In particular, for $Z = \phi$ we get $\bar{\tau}\phi : T_{\Sigma}\times Q \dot{\rightarrow} T_{\Omega}$. The last proposition establishes that to every top-down tree transformation corresponds a unique pure bottom-up transformation which in general is not a tree transformation anymore (the codomain is a certain function set and not a set of trees).

For (T,η,μ) and $(\bar{T},\bar{\eta},\bar{\mu})$ monads in a category $\underline{C}$ and $Q : \underline{C} \rightarrow \underline{C}$ a functor, an *inverse state transformation* of the above monads is a natural transformation $\bar{\tau} : QT \dot{\rightarrow} \bar{T}Q$ such that $\bar{\tau}\cdot Q\eta = \bar{\eta}Q$ and $\bar{\tau}\cdot Q\mu = \bar{\mu}Q\cdot\bar{T}\bar{\tau}\cdot\bar{\tau}T$.

Just as in the case of direct state transformations we have

Theorem The state transformation $\bar{\tau} : QX^{@} \dot{\rightarrow} \bar{T}Q$ on the free monad is an inverse state transformation $(X^{@},\eta,\mu) \rightarrow (\bar{T},\bar{\eta},\bar{\mu})$.

In the particular case of top-down tree transformations the above proposition corresponds to Thatcher [22], p. 353, lemma 6.7.

For $\bar{\tau} : QT \dot{\rightarrow} \bar{T}Q$ and $\bar{\rho} : SK \dot{\rightarrow} LS$ inverse state transformations define their composition $\bar{\rho}*\bar{\tau}$ as $\bar{\rho}Q\cdot S\bar{\tau}$.

Theorem The composition of inverse state transformations is an inverse state transformation.

For direct state transformations we can prove:

Theorem Given monads (T,η,μ) and $(\bar{T},\bar{\eta},\bar{\mu})$ in a category $\underline{C}$, there is a 1-1 correspondence between direct state transformations $\bar{\tau} : TQ \dot{\rightarrow} Q\bar{T}$ and liftings $\bar{Q} : \underline{C}^{\bar{T}} \rightarrow \underline{C}^{T}$ such that $U^{T}\bar{Q} = QU^{\bar{T}}$ where $U^{T} : \underline{C}^{T} \rightarrow \underline{C}$ and $U^{\bar{T}} : \underline{C}^{\bar{T}} \rightarrow \underline{C}$ are the forgetful functors.

For inverse state transformations only a weaker dual property is true. Denote with $\underline{C}_{T}$ the full subcategory of $\underline{C}^{T}$ consisting of free T-algebras only. Let F_{T} be the functor sending an object Q of $\underline{C}$ to the free T-algebra $(TQ,\mu Q)$ on Q generators. Then we have:

Theorem There exists a bijective correspondence between inverse state transformations $\bar{\tau} : QT \dot{\rightarrow} \bar{T}Q$ and liftings $\bar{Q} : \underline{C}_{T} \rightarrow \underline{C}_{\bar{T}}$ such that $F_{\bar{T}}\bar{Q} = QF_{T}$.

$\underline{C}_T$ is called the Kleisli category for the monad (T,η,μ) (cf. [15], p. 143). It is interesting to observe that the morphisms of $\underline{C}_T$ for the tree monad are maps of the form $Z \to T_{\Sigma,Y}$ studied in [22].

REFERENCES

[1] S. Alagić, Natural State Transformations, COINS Technical Report 73B-2, University of Massachusetts at Amherst, 1973.

[2] M. A. Arbib and E. G. Manes, Machines in a Category, SIAM Review, April, 1974.

[3] M. A. Arbib and E. G. Manes, The Monoid of a Machine in a Category, to appear.

[4] M. A. Arbib and E. G. Manes, Adjoint Machines, State Behaviour Machines, and Duality, Technical Report 73B-1, February, 1973, Computer and Information Science, University of Massachusetts, Amherst; to appear in J. Pure Appl. Alg.

[5] M. A. Arbib and E. G. Manes, Kleisli Machines, to appear.

[6] B. S. Baker, Tree Transductions and Families of Tree Languages, Proceedings of Fifth Annual ACM symposium on theory of computing, May, 1973.

[7] M. Barr, Coequalizers and Free Triples, Math. Z., 116 (1970), 307-322.

[8] J. Beck, Distributive Laws, Lecture Notes in Mathematics, Vol. 80, Springer-Verlag.

[9] E. J. Dubuc, Kan Extensions in Enriched Category Theory, Lecture Notes in Mathematics, Vol. 145, Springer-Verlag, 1970.

[10] J. Engelfriet, Bottomup and Topdown Treetransformations: A Comparison, Memorandum No. 19, July, 1971, Techniche Hogeschool Twente, Netherlands.

[11] S. Eilenberg and J. B. Wright, Automata in General Algebras, Information and Control, 11 (1967).

[12] S. Ginsburg, The Mathematical Theory of Context-Free Languages, McGraw-Hill, New York, 1966.

[13] G. Grätzer, Universal Algebra, D. Van Nostrand, Princeton, 1968.

[14] H. Kleisli, Every Standard Construction is Induced by a Pair of Adjoint Functors, Proc. Am. Math. Society, 16 (1965).

[15] S. Mac Lane, Categories for the Working Mathematician, Springer-Verlag, 1972.

[16] E. G. Manes, A Triple Miscellany: Some Aspects of the Theory of Algebras Over a Triple, Thesis, Wesleyan University, 1967.

[17] E. G. Manes, Algebraic Theories, Springer-Verlag, to appear.

[18] J. P. Meyer, Induced Functors on Categories of Algebras, The John Hopkins University, Baltimore, 1972, preprint.

[19] J. Mezei and J. B. Wright, Generalized Algol-Like Languages, Information and Control, 11 (1967).

[20] W. C. Rounds, Mappings and Grammars on Trees, Mathematical System Theory, 4 (1970).

[21] J. W. Thatcher, Characterizing Derivation Trees of Context-Free Grammars Through a Generalization of Finite Automata Theory, Journal of Computer and System Sciences, 1 (1967).

[22] J. W. Thatcher, Generalized^2 Sequential Machine Maps, Journal of Computer and System Sciences, 4 (1970).

[23] J. W. Thatcher, There's a Lot More to Finite Automata Theory Than You Would Have Thought, RC 2852, 1970, IBM, Yorktown Heights.

REALIZATION OF MULTILINEAR AND[1] MULTIDECOMPOSABLE MACHINES

Brian D.O. Anderson
Electrical Engineering
University of Newcastle
NSW, Australia

Michael A. Arbib
Computer and Information Science

Ernest G. Manes
Mathematics

University of Massachusetts
Amherst, Massachusetts 01002, U.S.A

INTRODUCTION

In [1], there appears a category theoretic view of linear systems over a vector space which at the same time extends to a number of other classes of systems, for example, group machines. Our object here is to present some results on multilinear systems in the same vein as [1]. For most of the paper, we shall pose the results in terms of vector spaces. However, at the end of the paper, we indicate generalizations applicable to a category more general than Vect.

An outline of our results is as follows: we define a K-line multilinear machine in terms of linear, bilinear,...(K-1)-line multilinear machines via an internal description, and we show that the associated input-output map is multilinear. Then we tackle the converse problem: passage from a prescribed multilinear map to an internal description. We discuss questions of the reachability, observability, and minimality of realizations.

[1] The research reported in this paper was supported in part by Grant #GJ35759 of the National Science Foundation and the Australian Research Grants Committee.

DEFINITION OF A MULTILINEAR SYSTEM

First, we define a linear system as a sextuple (Q,I,Y,F,G,H) where Q, I,Y are the state, input and output (vector) spaces, and $F:Q\to Q$, $G:I\to Q$, $H:Q\to Y$ are linear transformations; the system operates according to

$$\begin{aligned} q(k+1) &= Fq(k) + Gu(k) \\ y(k) &= Hq(k) \end{aligned} \qquad (1)$$

where u,q,y are in I,Q,Y respectively.

In diagram form we have

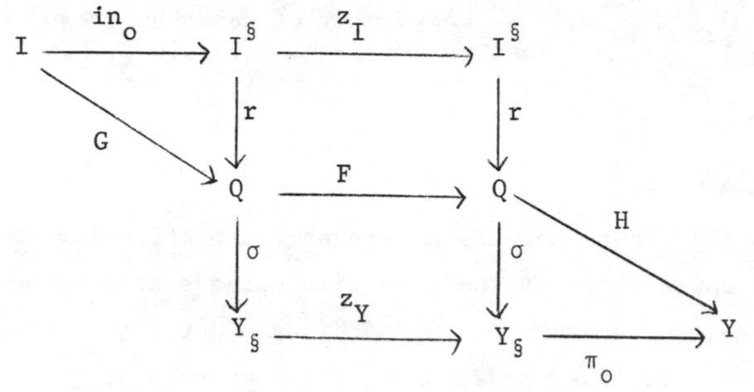

(2)

Here, $I^§$ denotes the countably infinite copower of I, which one can identify with left infinite sequences of finite support $(\ldots,0,\ldots,0,i_k,i_{k-1},\ldots,i_o)$ with the injection $in_j: I\to I^§: i_j \mapsto (\ldots,0,\ldots,0,i_j,0,\ldots,0)$ into the (j+1) -th slot from the right; one can think of i_ℓ as $u(-\ell)$. Also, $Y_§$ denotes the countably infinite power of Y, which one can identify with right infinite sequences $(y_o,y_1,y_2\ldots)$ thinking of y_ℓ as $y(\ell+1)$. One has $\pi_j: Y_§ \to Y: (y_o,y_1,y_2\ldots)\to y_j$. The morphisms z_I and z_Y are shift operators, defined respectively by $z_i in_j = in_{j+1}$ and $\pi_{j+1} = \pi_j z_Y$; z_I has the effect of left shift with addition of zero in the zero position and z_Y of left shift, discarding the element in the zero position prior to the shift. The maps r and σ are the reachability map (mapping past input sequences into the present state) and the observability map (mapping the present state into a future output sequence, assuming zero future inputs). For a full discussion, see [1].

A multilinear system with K input lines numbered 1, 2,...,K is defined by an equation of the form

$$q(k+1) = Fq(k) + \sum \varepsilon_{i_{j+1}\cdots i_K}[y_{i_1\cdots i_j}(k)\otimes u_{i_{j+1}}(k)\otimes\cdots\otimes u_{i_K}(k)]+G(u_1(k)\otimes\cdots\otimes u_K(k))$$

$$y(k) = Hq(k) \tag{3}$$

where $I_1,\ldots,I_K$ are K input spaces, Q is the state space, Y is the output space; $u_j \varepsilon I_j$, $q\varepsilon Q$, $y\varepsilon Y$; $y_{i_1\cdots i_j}(k)$ is the output of a known j-line multilinear system with input spaces $I_{i_1},\ldots,I_{i_j}$, the summation is over every partition of $(1,2,\ldots,K)$ into disjoint nonempty sets $(i_1,\ldots,i_j)$, $(i_{j+1},\ldots,i_K)$; $y_{i_1\cdots i_j}(k)\ \varepsilon\ \mathrm{Hom}(I_{i_{j+1}}\otimes\cdots\otimes I_{i_K},Q)$ and $\varepsilon_{i_{j+1}\cdots i_K}$ is the canonical evaluation map; and F,G,H are linear maps with obvious domain and codomain.

Figure 1 illustrates the idea for K=2 while Figure 2 illustrates K=3

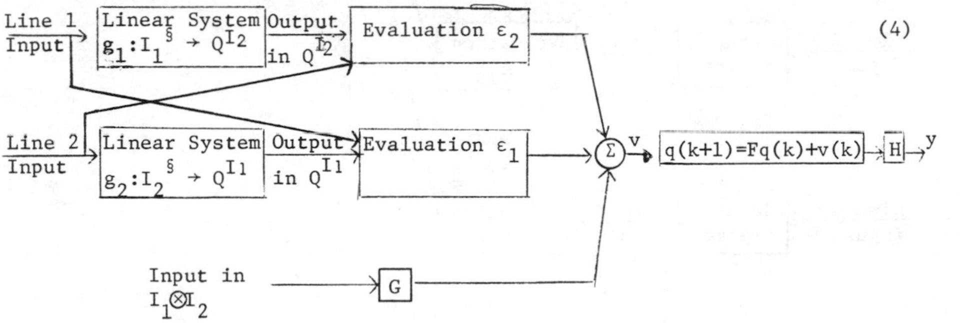

(4)

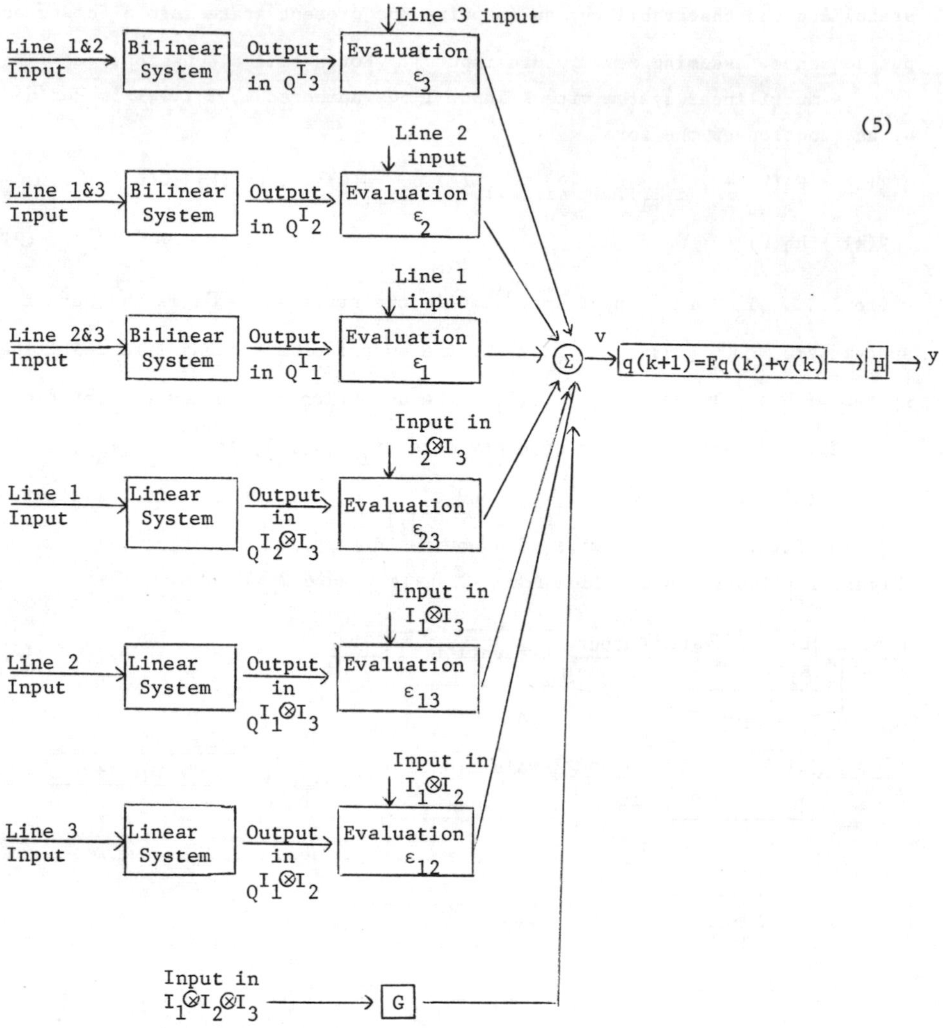

(5)

Figure 2

Each bilinear system has its own breakdown into linear systems.

OBTAINING INPUT-OUTPUT DESCRIPTION FROM INTERNAL DESCRIPTION

Theorem 1 A K-line multilinear system defines a linear map $f: I_1^{\S} \otimes I_2^{\S} \ldots \otimes I_K^{\S} \to Y$ which can be extended to a map $f^{\blacktriangle}: I_1^{\S} \otimes \ldots \otimes I_K^{\S} \to Y_{\S}$

such that $f^{\blacktriangle}$ possesses the dynamorphic property $f^{\blacktriangle}(z_{I_1} \otimes \ldots \otimes z_{I_K}) = z_Y f^{\blacktriangle}$, and $\pi_0 f^{\blacktriangle} = f$.

Remark One can also work with a multilinear map $\tilde{f} : I_1^{\S} \otimes \ldots \otimes I_K^{\S} \to Y$ which is associated with f in a canonical way.

In the bilinear case, one defines $f^{\blacktriangle}$ as $\sigma \cdot r$ from the following diagram (showing only one of the linear systems of (4)).

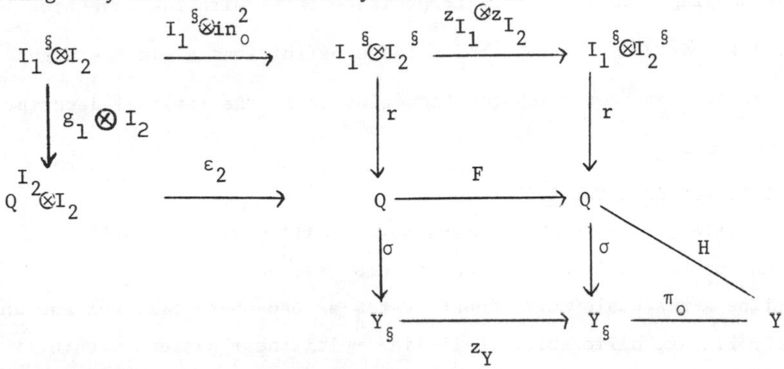

(6)

The map σ is constructed as for linear systems in [1]. Notice that given the definition (3) is completely equivalent. A set-up like that of (4) has been analyzed in [2-4], with the structure obtained via an analysis of the Nerode equivalence relation for a bilinear map $\tilde{f}: I_1^{\S} \times I_2^{\S} \to Y$. Let us now see how this structure will arise in an apparently quite different way, by generalizing some ideas of [1].

OBTAINING INTERNAL DESCRIPTION FROM INPUT-OUTPUT DESCRIPTION

Suppose $f : I_1^{\S} \otimes \ldots \otimes I_K^{\S} \to Y$; with no attention being paid to reachability or observability. An internal description is obtained as follows: take $Q = Y_{\S}$, $r = f^{\blacktriangle}$, $H = \pi_0$, $F = z_Y$

$$
\begin{array}{ccc}
I_1^{\S} \otimes \cdot \cdot \otimes I_K^{\S} & \xrightarrow{z_{I_1} \otimes \ldots \otimes z_{I_K}} & I_1^{\S} \otimes \cdot \cdot \otimes I_K^{\S} \\
\downarrow{f^{\blacktriangle}} & & \downarrow{f^{\blacktriangle}} \\
Y_{\S} & \xrightarrow[z_Y]{} & Y_{\S} \xrightarrow[\pi_0]{} Y
\end{array}
$$

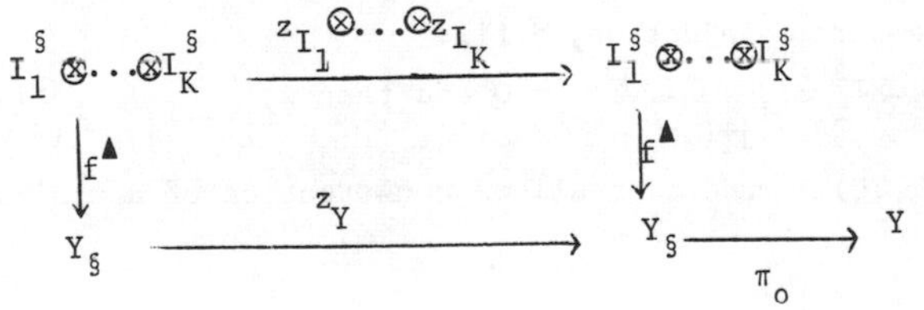

(7)

By using the appropriate zero injection, $I_{i_1}^{\S} \otimes \ldots \otimes I_{i_j}^{\S} \otimes I_{i_{j+1}} \otimes \ldots \otimes I_{i_K} \to I_1^{\S} \otimes \ldots \otimes I_K^{\S}$, obtain a map $g^{\bullet}_{i_1 \ldots i_j} : I_{i_1}^{\S} \otimes \ldots \otimes I_{i_j}^{\S} \otimes I_{i_{j+1}} \otimes \ldots \otimes I_{i_K} \to Y_{\S}$. Then $g_{i_1 \ldots i_j} : I_1^{\S} \otimes \ldots \otimes I_{i_j}^{\S} \to \mathrm{Hom}(I_{i_{j+1}} \otimes \ldots \otimes I_{i_K}, Y_{\S})$ such that $\varepsilon_{i_{j+1} \ldots i_K}(g_{i_1 \ldots i_j} \otimes I_{i_{j+1}} \otimes \ldots \otimes I_{i_K}) = g^{\bullet}_{i_1 \ldots i_j}$ is uniquely definable.

The mapping G is defined as composition of $f^{\blacktriangle}$ with the injection $in_o^1 \otimes \ldots \otimes in_o^K : I_1 \otimes \ldots \otimes I_K \to I_1^{\S} \otimes \ldots \otimes I_K^{\S}$; these definitions yield the right commutative diagrom from which (by induction on K) the internal description can be recovered.

REACHABILITY AND OBSERVABILITY

Definitions of reachability and observability are obtainable via induction on the number of lines of a multilinear system.

Define a reachable multilinear system as one where r is epi and in case $K \geq 1$, the linear, bilinear,...(K-1)-line multilinear systems within it are also reachable. (Notice that if the second condition fails, the first need not; it is only the input-output properties of the linear,...(K-1)-line systems which can affect r, see (4) or (6)).

An observable system is one where σ is mono, and the linear, bilinear... (K-1)-line multilinear systems within it are also observable.

MINIMAL REALIZATION

A minimal realization of a map $f : I_1^{\S} \otimes \ldots \otimes I_K^{\S} \to Y$ is one which is reachable and observable.

Such a realization can be obtained as follows. Factor $f^{\blacktriangle}$ as σ r where r is epi, σ is mono; take $Q = r(I_1^{\S} \otimes \ldots \otimes I_k^{\S})$ (Effectively, $Q = \mathrm{Im}(f^{\blacktriangle})$. Using the appropriate zero injection, define
$g^{\bullet}_{i_1 \ldots i_j} : I_{i_1}^{\S} \otimes \ldots \otimes I_{i_j}^{\S} \otimes I_{i_{j+1}} \otimes \ldots \otimes I_{i_K} \to Q$ and then $g_{i_1 \ldots i_j} : I_{i_1}^{\S} \otimes \ldots \otimes I_{i_j}^{\S} \to \mathrm{Hom}(I_{i_{j+1}} \otimes \ldots \otimes I_{i_K}, Q)$. Induction allows construction of a minimal realization for each such $g_{i_1 \ldots i_j}$.

Define the <u>state-space set</u> $\mathcal{Q}$ of a K-line multilinear system by: $Q \in \mathcal{Q}$, and elements of the state-space set of all linear, bilinear,...(K-1)-linear multilinear system within the K-line multilinear system are in $\mathcal{Q}$. Then we have the expected result:

<u>Theorem</u> The elements, appropriately ordered, of the state-space sets of two minimal realizations of the same $f = I_1^{\S} \otimes \ldots \otimes I_K^{\S} \rightarrow Y$ are isomorphic.

CATEGORIES OTHER THAN VECT

Let K be a symmetric closed strict monoidal category with countable coproducts and products. One can consider diagrams such as (5) and observe they define, via a generalized sort of recursion, a morphism $f = I_1^{\S} \otimes \ldots \otimes I_K^{\S} \rightarrow Y$, and one can start with such a morphism and show that there exist diagrams of the form (5) which define realizations of f. (One uses the adjointness property of the tensor product to pass between $\dot{g}_{i_1 \ldots i_j}$ and $g_{i_1 \ldots i_j}$ in the construction, which otherwise is the same mutatis mutandis, as for <u>Vect</u>). If $\mathcal{E}$-$\mathcal{M}$ factorizations exist in K, one can define reachable and observable realizations of a prescribed f. By analogy with [1], we might call these systems <u>multidecomposable</u>.

REFERENCES

1. M.A. Arbib and E.G. Manes, "Foundations of System Theory:Decomposable Systems", <u>Automatica</u>, May 1974.

2. M.A. Arbib, "A Characterization of Multilinear Systems", <u>IEEE Transaction on Automatic Control</u>, Vol. AC-14, No. 6, Dec. 1969, pp. 699-702.

3. G. Marchesini and G. Picci, "Some Results on the Abstract Realization Theory of Multilinear Systems", in <u>Theory and Applications of Variable Structure Systems</u>, ed. R.R. Mohler and A. Ruberti, Academic Press, New York, 1972, pp. 109-135.

4. E. Fornassini and G. Marchesini, "On the Internal Structure of Bilinear Input-Output Maps", <u>Proceedings of NATO Conference on Control and System Theory</u>, Imperial College, London, 1973.

5. R.E. Kalman, P.L. Falb and M.A. Arbib, <u>Topics in Mathematical System Theory</u>, McGraw Hill Book Co., New York, 1969.

FUZZY MORPHISHMS IN AUTOMATA THEORY[1]

M. A. Arbib
Computer and Information Science

E. G. Manes
Mathematics

University of Massachusetts
Amherst, MA 01002 USA

A function $f : A \longrightarrow B$ generalizes to a "fuzzy function" $F : A \longrightarrow B$ in a number of ways. F can be a partial function, a relation, a stochastic matrix, a fuzzy relation $A \times B \longrightarrow [0,1]$ ([11]) or an L-fuzzy relation $A \times B \longrightarrow L$ ([7]). In all cases, a composition $F : A \longrightarrow B$, $G : B \longrightarrow C \quad \longmapsto \quad F \circ G : A \longrightarrow C$ is defined, giving rise to a category of fuzzy functions containing functions as a subcategory. In this paper we axiomatize fuzzy functions so as to substantially generalize the above examples while meeting the needs of a theory of non-deterministic automata. Since the abstract theory is formally equivalent to the theory of monads ([8, chapter XI]), that is, to the theory of algebraic theories ([9]), new examples arise and the concepts of "decider" and "decider homomorphism", currently disguised in more traditional notation in the motivating examples, emerge clearly. The theory of machines in a category ([1] and its sequels) generalizes at once to a non-deterministic form. As was discovered independently by Elizabeth Burroni ([4]), this theory is closely connected to Beck distributive laws ([3]). When the base category is closed, our

[1]The research reported in this paper was supported in part by the National Science Foundation under Grant No. GJ 35759.

theory is much like that of Ehrig's ([5, Kapitel 6] , [6]); however, rather than attempting to recapture the pathological "nonexistence of minimal realizations" (a dogma originating with Starke [10]), we use the theory of composite algebras to present an algebraic generalization of non-deterministic automata that admits a minimal realization theory which in the classical case ([10, Part II]) yields the slogan: the minimal realization of a non-deterministic automaton is an L-set ([7]) for some L.

A <u>fuzzy theory</u> in a category $\mathcal{K}$ is a triple $\underset{\sim}{T} = (T, \circ, e)$ as follows. T assigns to each object A a new object AT of "fuzzy states". A <u>fuzzy morphism</u> $F : A \longrightarrow B$ is a morphism $F : A \longrightarrow BT$. A composition of fuzzy morphisms $F : A \longrightarrow B,\ G : B \longrightarrow C \ \mapsto\ F \circ G : A \longrightarrow C$ is assigned by $\circ$. e assigns to each object A a morphism $Ae : A \longrightarrow AT$ ("pure states"). The three axioms imposed are:

(1) $(F \circ G) \circ H = F \circ (G \circ H)$

(2) $f^{\phi} \circ G = f.G$ for all $f : A \longrightarrow B,\ G : B \longrightarrow C$ (where "." denotes composition in $\mathcal{K}$, and $f^{\phi} : A \longrightarrow B$ is f considered fuzzy via $f^{\phi} = f.Be$.

(3) $F \circ Be = F$

The resulting category $\mathcal{K}(\underset{\sim}{T})$ of $\mathcal{K}$-objects and fuzzy morphisms admits $(\cdot)^{\phi}$ as a functor $\mathcal{K} \longrightarrow \mathcal{K}(\underset{\sim}{T})$. In examples 4-7, $\mathcal{K}$ = <u>Set</u>.

(4) $AT = A + \{u\}$ ("u" for "undefined"), $\circ$ is usual composition of partial functions, e is the obvious inclusion.

(5) $AT = \{S : S \subset A\}$, $\circ$ is the usual composition of relations, e is the "singleton map". (Similarly, one may use nonempty subsets, finite subsets, etc.)

(6) $AT = \{p : p$ is a function from A to $[0,1]$ such that p has finite support and $\sum p(a) = 1\}$ = the set of probability distributions on A, $\circ$ is the usual composition of stochastic matrices, Ae sends a to the unique p with $p(a) = 1$.

(7) Let $(L,*)$ be a complete lattice ordered semigroup ([7, p. 154]). Define $AT = L^A$, $\circ$ as in [7, p. 161] and e by using characteristic functions of singletons.

(8) <u>Contraction principle.</u> $id_{AT} : AT \longrightarrow AT$ is a fuzzy morphism $id_{AT} : AT \longrightarrow A$. In (5), $id_{AT} : AT \longrightarrow A = \{(S,a) : a \in S\}$.

(9) <u>Asymmetry theorem.</u> Given $F : A \longrightarrow B$, $G : B \longrightarrow C$ then $F \circ G = F.G^{\#}$ where $G^{\#} : BT \longrightarrow CT = id_{BT} \circ G \circ id_{CT}$.

(10) <u>Characterization theorem.</u> The passage $(T, \circ, e) \longmapsto (T, e, m)$ where $Am : ATT \rightarrow AT = id_{ATT} \circ id_{AT}$ and T acts as a functor via $f : A \rightarrow B$ $\longmapsto$ $fT : AT \longrightarrow BT = id_{AT} \circ f^{\not c}$, extablishes a bijective correspondence between fuzzy theories in $\mathcal{K}$ and monads in $\mathcal{K}$ ([8, chapter VI]). The reverse passage is well known as the "Kleisli category" construction.

The characterization theorem allows a universal-algebraic interpretation of fuzzy theories: AT = "formulas with variables in A", $\circ$ = "clone composition, i.e. substitution of formulas for variables in formulas" and e = "variables qua formulas". Moreover, the monad point of view suggest the concept of a <u>T-algebra,</u> being a pair (A, ξ) where $\xi : AT \longrightarrow A$ satisfies

$$\begin{array}{ccc} A & \xrightarrow{Ae} & AT \\ & {\scriptstyle id_A} \searrow & \downarrow \xi \\ & & A \end{array} \qquad \begin{array}{ccc} ATT & \xrightarrow{\xi T} & AT \\ {\scriptstyle Am} \downarrow & & \downarrow \xi \\ AT & \xrightarrow{\xi} & A \end{array} \qquad (11)$$

A <u>T-homomorphism</u> $f : (A, \xi) \longrightarrow (A', \xi')$ satisfies

$$\begin{array}{ccc} AT & \xrightarrow{fT} & A'T \\ \xi \downarrow & & \downarrow \xi' \\ A & \xrightarrow{f} & A' \end{array} \qquad (12)$$

(13) Let AT be the free monoid A^* with "inclusion of the generators" $Ae : A \rightarrow AT$ ("words of length 1"). If (M,m) is a monoid and $f : A \longrightarrow M$ is a function then there exists a unique monoid homomorphism $f^{\#} : AT \longrightarrow (M,m)$ extending f, namely $f^{\#} : a_1 \cdots a_n \longmapsto a_1 f \cdots a_n f$. Defining $F \circ G = F.G^{\#}$ (see (9)), $(T, \circ, e)$ is a fuzzy theory of sets. A T-algebra <u>is</u> a monoid (ξ realizes formal multiplication; conversely, define $xy = (xy)\xi$) and a T-homomorphism is a monoid homomorphism.

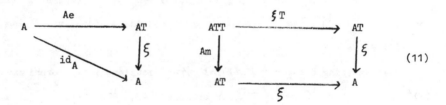

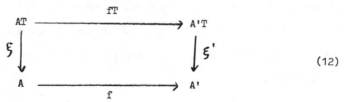

The construction of (13) generalizes to any equationally defineable class such as groups, rings, lattices or Boolean σ-rings (infinitary operations are permissible!).

From the point of view of fuzzy theories, a $\underset{\sim}{T}$-algebra is a <u>decider</u> (with <u>decision map</u> ξ, choosing a pure state for each fuzzy one according to compatibility laws (11)). For $\underset{\sim}{T}$ as in (5), a decider is a complete lattice (ξ is the supremum map) and $\underset{\sim}{T}$-homomorphisms are supremum preserving. In the context of (4), (A,ξ) is a set with base point and homomorphisms preserve the base point. With respect to (6), a decider is a generalized convex set with ξ assigning "center of gravity"; since Linton (unpublished) has shown the existence of 2-element models, not all deciders are convex subsets of a real linear space.

There are examples motivated neither by fuzzy theories nor by universal algebra. For instance:

(14) Let $AT = \{\mathcal{U} : \mathcal{U}$ is an ultrafilter on $A\}$, $Ae =$ "principal ultrafilter map", $G^{\#} : BT \longrightarrow CT$ via

$$\mathcal{U} \in BT \longmapsto \bigcup_{U \in \mathcal{U}} \bigcap_{u \in U} G_u \in CT$$

$\underset{\sim}{T}$-algebras are compact Hausdorff spaces (ξ is convergence) and $\underset{\sim}{T}$-homomorphisms are the continuous maps.

There are also many examples in categories other than <u>Set</u>.

Let $X : \mathcal{K} \longrightarrow \mathcal{K}$ be an input process (see [1] and part 3 of the introduction to this volume) and let (T,σ,e) be a fuzzy theory in $\mathcal{K}$. Since the dynamics of a deterministic X-machine is a morphism $\delta : QX \longrightarrow Q$ one expects the dynamics of a non-deterministic --more precisely, a $\underset{\sim}{T}$-deterministic-- X-machine to be a fuzzy morphism of form $\Delta : QX \longrightarrow Q$. With the desiratum that a $\underset{\sim}{T}$-deterministic X-machine in $\mathcal{K}$ be "run" as a deterministic machine of some sort in $\mathcal{K}(\underset{\sim}{T})$ we hope for a lifting

$$\begin{array}{ccc} \mathcal{K}(\underset{\sim}{T}) & \overset{\overline{X}}{- - - - - \rightarrow} & \mathcal{K}(\underset{\sim}{T}) \\ {\scriptstyle (\cdot)^{\natural}}\uparrow & & \uparrow{\scriptstyle (\cdot)^{\natural}} \\ \mathcal{K} & \xrightarrow[X]{\qquad\qquad} & \mathcal{K} \end{array} \qquad (15)$$

In general, such liftings need not exist nor need they be unique. However,

(16) Each lift $\overline{X}$ as in (15) is an input process in $\mathcal{K}(\underset{\sim}{T})$ and, moreover, the structure of the free dynamics ([1]) over an object is independent of the choice of lift.

Notice that $(id_{AT} : AT \longrightarrow A)\overline{X}$ is a fuzzy map of form $ATX \longrightarrow AX$, that is, a $\mathcal{K}$-morphism $A\lambda : ATX \longrightarrow AXT$. It is easy to prove that $\overline{X}$ is determined by λ so that suitable axioms on λ characterize $\overline{X}$. We have

(17) (Done independently in [4]). Liftings $\overline{X}$ as in (15) are in bijective correspondence with Beck distributive laws ([3]) $\lambda : TX \longrightarrow XT$.

The distributive law λ gives rise to a monad structure on $X^{@}T$ ([3]) whose algebras --the X-$\underset{\sim}{T}$ <u>composite algebras</u>-- may be identified as those (A, δ, ξ) with (A, δ) an X-dynamics and (A, ξ) a $\underset{\sim}{T}$-algebra, subject to the <u>composite law</u>

$$\begin{array}{ccccc} ATX & \xrightarrow{A\lambda} & AXT & \xrightarrow{\delta T} & AT \\ {\scriptstyle \xi X}\downarrow & & & & \downarrow{\scriptstyle \xi} \\ AX & & \xrightarrow[\delta]{\qquad\qquad\qquad} & & A \end{array} \qquad (18)$$

In the adjoint machine case (see [2] and part 3 of the introduction to this volume) we may attempt to "run" a $\underset{\sim}{T}$-deterministic machine $\Delta : QX \longrightarrow QT$ with initial state $\tau : I \longrightarrow QT$ and output $\beta : Q \longrightarrow YT$ as follows. $IX^{@}T$ is a composite algebra (the free one on I generators). QT is a composite algebra with X-dynamics

$$QTX \xrightarrow{QT\lambda} QXT \xrightarrow{\Delta^{\#}} QT$$

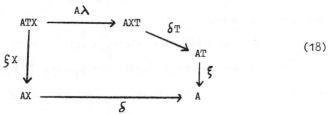

Moreover, $YX_@T$ is a composite algebra, the X-dynamical structure being

$$YX_@TX \xrightarrow{YX_@\lambda} YX_@XT \xrightarrow{LT} YX_@T$$

Consult diagram (19). The <u>reachability map</u> r is the unique X-$\underset{\sim}{T}$-homomorphic extension of τ whereas the <u>observability map</u> σ is the X-dynamorphic extension of $\beta^{\#}$ which is provably a $\underset{\sim}{T}$-homomorphism. The <u>response</u> is the X-$\underset{\sim}{T}$-homomorphism $f = r\sigma : IX^@T \longrightarrow YTX_@$. The minimal realization is

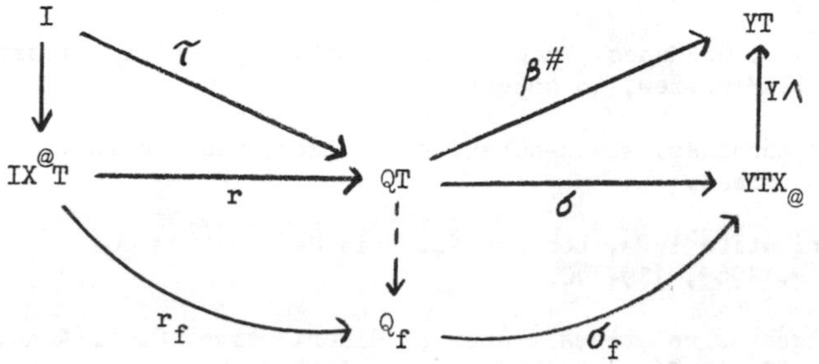

(19)

is then built (cf. [2]) on the $\mathcal{E}$-$\mathcal{M}$ factorization $r_f\sigma_f$ of f as shown in (19); under reasonable circumstances (including the case $\mathcal{K}$ = <u>Set</u>) Q_f is a composite subalgebra. This is another example of a minimal realization theory for a monadic functor $U : \mathcal{A} \longrightarrow \mathcal{B}$ with (left and) right adjoints without $\mathcal{A}$ having the form Dyn(X) (cf. the remarks made in part 3 of the introduction to this volume); here, $\mathcal{A}$ is the category of X-$\underset{\sim}{T}$ composite algebras, $\mathcal{B}$ is the category of $\underset{\sim}{T}$-algebras and U is the forgetful functor $(A, \delta, \xi) \longmapsto (A, \xi)$.

When $\mathcal{K}$ is <u>Set</u> and $X = - \times X_o$: <u>Set</u> $\longrightarrow$ <u>Set</u>, for any fuzzy theory $\underset{\sim}{T}$ there is a canonical distributive law $A\lambda : AT \times X_o \longrightarrow (A \times X_o)T$ defined by sending (s,x) to the value of s under in_xT where in_x sends a in A to (a,x) in $A \times X_o$. Arbitrary (X-$\underset{\sim}{T}$-homomorphic) responses $f : IX^@T \longrightarrow YTX_@$ --which are always in bijective correspondence with $\mathcal{K}$-morphisms $I \longrightarrow YTX_@$-- are then just fuzzy morphisms $X_o^* \longrightarrow Y$ when I = 1. For $\underset{\sim}{T}$ as in (5), each $\underset{\sim}{T}$-deterministic machine with Q finite has a finite minimal realization Q_f which is a complete lattice on which X_o acts by supremum preserving endomorphisms. Via the mapping $Q \longrightarrow QT \longrightarrow Q_f$, the minimal realization of a classical nondeterministic automaton is an L-set ([7]) with $L = Q_f$.

REFERENCES

1. M. A. Arbib and E. G. Manes, Machines in a category: an expository introduction, SIAM Review, to appear.

2. _____ Adjoint machines, state-behavior machines, and duality, Automatica, to appear.

3. J. Beck, Distributive laws, Lecture Notes in Mathematics 80, Springer-Verlag, 1969, 119-140.

4. E. Burroni, Algèbres relatives à une loi distributive, C. R. Acad. Sci. Paris, t. 276 (5 février 1973), Serie A, 443-446.

5. H. Ehrig, K. D. Kiermeier, H.-J. Kreowski, W. Kühnel, Systematisierung der Automatentheorie: Seminarbericht 1972/73, Bericht Nr. 73-08, Technische Universität Berlin, Fachbereich 20 - Kybernetik, April, 1973.

6. H. Ehrig, H.-J. Kreowski, Power and initial automata in pseudoclosed categories, these proceedings.

7. J. A. Goguen, L-fuzzy sets, J. Math. Anal. Appl. 18, 145-174, 1967.

8. S. Mac Lane, Categories for the Working Mathematician, Springer-Verlag, 1972.

9. E. G. Manes, Algebraic Theories, Springer-Verlag, to appear.

10. P. H. Starke, Abstract Automata, North-Holland, 1972; translation of Abstrakte Automaten, VEB, Berlin, 1969.

11. L. A. Zadeh, Fuzzy sets, Inf. Cont. 8, 1965, 338-353.

TIME-VARYING SYSTEMS[1]

Michael A. Arbib
Computer and Information Science

Ernest G. Manes
Mathematics

University of Massachusetts
Amherst, Massachusetts 01002, U.S.A.

We show that we can model time-varying systems of adjoint processes in a category $\mathcal{K}$ as normal adjoint processes in the category $\mathcal{K}^{\underline{Z}}$. We saw the richness of adjoint machines in [3]--with adjoint machines including sequential machines, nondeterministic machines, Boolean machines, metric machines and topological machines. In "Machines in a Category" [1] we gave the category theorist's definition of functors and of left adjoints, and said that the input structure of a machine should not be regarded as a set of applicable inputs, but rather as a process which transforms the state-space Q into a new object QX on which the dynamics can act:

1. DEFINITION: Given a functor $X: \mathcal{K} \longrightarrow \mathcal{K}$, $\mathrm{Dyn}(X)$ denotes the category of X-dynamics whose objects are pairs (Q,δ), where Q is a $\mathcal{K}$-object and $\delta: QX \longrightarrow Q$ is a $\mathcal{K}$-morphism; while dynamorphisms

[1] The research reported in this paper was supported in part by the National Science Foundation under Grant No. GJ 35759.

$g: (Q,\delta) \longrightarrow (Q',\delta')$ are $\mathcal{K}$-morphisms $g: Q \longrightarrow Q'$ for which the diagram

$$\begin{array}{ccc} QX & \xrightarrow{\delta} & Q \\ {\scriptstyle gX}\downarrow & & \downarrow{\scriptstyle g} \\ Q'X & \xrightarrow{\delta'} & Q' \end{array}$$

commutes.

We then said that for X to be interesting, it must be an input process in the following sense:

2. DEFINITION: X is an input process if the forgetful functor $\mathrm{Dyn}(X) \longrightarrow \mathcal{K}: (Q,\delta) \mapsto Q$ has a left adjoint; i.e. if for each $Q \in \mathcal{K}$ there exists a free dynamics $\mu_o: (QX^@)X \to QX^@$ with a $\mathcal{K}$-morphism $\eta: Q \to QX^@$ such that given any X-dynamics (Q',δ') and any $\mathcal{K}$-morphism $f: Q \to Q'$, there exists a unique dynamorphism $\psi: (QX^@,\mu_o) \to (Q',\delta')$ such that $\psi \cdot \eta = f$:

$$\begin{array}{ccc} Q & \xrightarrow{\eta} & QX^@ \\ & {\scriptstyle f}\searrow & \downarrow{\scriptstyle \psi} \\ & & Q' \end{array} \qquad\qquad \begin{array}{ccc} (QX^@)X & \xrightarrow{\mu_o} & QX^@ \\ {\scriptstyle \psi X}\downarrow & & \downarrow{\scriptstyle \psi} \\ Q'X & \xrightarrow{\delta'} & Q' \end{array}$$

We assume that the reader knows (or can look up in Mac Lane [5]) the categorical definition of coproducts. We shall next define what it means for a functor to have a right adjoint, and establish broad conditions under which an $X: \mathcal{K} \to \mathcal{K}$ with right adjoint will be an input process:

3. DEFINITION: A functor $F: \mathcal{A} \to \mathcal{B}$ has a right adjoint if there exists a functor $F^{\bullet}: \mathcal{B} \to \mathcal{A}$ (the right adjoint of F) such that to each B in there corresponds a $\mathcal{B}$-morphism $BFF^{\bullet} \xrightarrow{\varepsilon} B$ such that to each $\mathcal{B}$-morphism $g: B'F \to B$ there corresponds a unique $\mathcal{A}$-morphism $\phi: B' \to BF^{\bullet}$ such that

$$\begin{array}{ccc} B & \xleftarrow{\varepsilon} & BF^{\bullet}F \\ & {\scriptstyle g}\nwarrow & \uparrow{\scriptstyle \phi F} \\ & & B'F \end{array} \qquad\qquad \begin{array}{c} BF^{\bullet} \\ \uparrow{\scriptstyle \phi} \\ B' \end{array} \tag{1}$$

commutes.

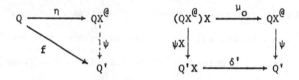

4. EXAMPLE: Let Vect = <Vector Spaces and Linear Maps>, and let $X: \underline{Vect} \to \underline{Vect}$ be the identity functor ($Q \mapsto Q$; $f \mapsto f$). Then an X-dynamics is just a linear map $F: QX = Q \to Q$, as in the linear time invariant systems of [2]. Clearly X is its own right adjoint--setting $F = F^{\bullet} = X$, $\varepsilon = id_B$, we have that (1) is satisfied with $\phi = g$:

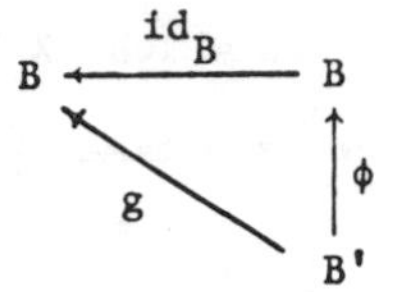

5. EXAMPLE: Let Set = <Sets and Maps>, and let $X = -\times X_o: \underline{Set} \to \underline{Set}$ be the functor $Q \mapsto Q\times X_o$; $f \mapsto f\times X_o$ where

$$f\times X_o: Q\times X_o \longrightarrow Q'\times X_o: (q,x) \mapsto (f(q),x) .$$

Then an X-dynamics is just a map $\delta: Q\times X_o \longrightarrow Q$, the next-state function of a sequential machine. $X = -\times X_o$ has right adjoint $(-)^{X_o}$ which sends Q to the set Q^{X_o} of all maps from X_o to Q. $\varepsilon: B^{X_o} \times X_o \longrightarrow B$ is the evaluation $(f,x) \mapsto f(x)$, and we have that (1) is satisfied on taking $\phi(b'): X_o \longrightarrow B: x \mapsto g(b',x)$:

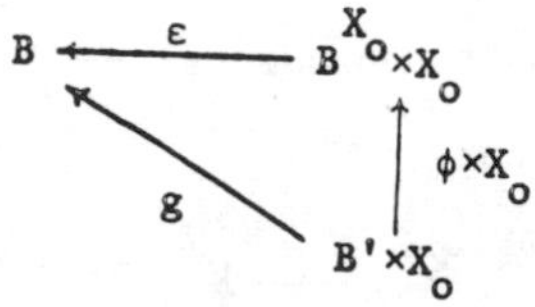

6. LEMMA: If $F: \mathcal{A} \to \mathcal{B}$ has a right adjoint, then F preserves coproducts. □

Now, given any functor $X: \mathcal{K} \to \mathcal{K}$, define X^n for any $n\geq 0$ by the rules

$$X^o = id$$

$$X^{n+1} = X^n \cdot X \qquad \text{for } n\geq 0$$

and then set X^* to be the countable coproduct

$$X^* = \coprod_{n\geq 0} X^n .$$

7. THEOREM: Let $\mathcal{K}$ have, and let $X: \mathcal{K} \to \mathcal{K}$ preserve, countable coproducts. Then X is an input process and $X^@ = X^*$. □

8. COROLLARY: If $\mathcal{K}$ has countable coproducts and $X: \mathcal{K} \to \mathcal{K}$ has a right adjoint, then X is an input process, and $X^@ = X^*$. We say such an X is an <u>adjoint process</u>. □

We now show that if X is an adjoint process in $\mathcal{K}$, then X induces an adjoint process $\overline{X}$ which yields "time-varying" X-dynamics. The construction of the category $\mathcal{K}^{\underline{Z}}$ in which $\overline{X}$ lives is clearly motivated by considering a state-space for each time $k \in \underline{Z}$:

$\mathcal{K}^{\underline{Z}}$-Objects: Sequences $Q = (Q_k \mid k \in \underline{Z})$ with each Q_k in $\mathcal{K}$.

$\mathcal{K}^{\underline{Z}}$-Morphisms: $f: Q \to Q' = (f_k: Q_k \to Q'_k \mid k \in \underline{Z})$ with each f_k in $\mathcal{K}$.

$\mathcal{K}^{\underline{Z}}$ is a category with $(f \cdot g)_k = f_k \cdot g_k$; $(id_Q)_k = (id_{Q_k})$.

[Note: The underlying scheme could be far more complex than $\underline{Z}$. We shall develop the implications of this observation elsewhere. See [5] for a discussion of functor categories.]

9. THEOREM: Let $X: \mathcal{K} \to \mathcal{K}$ where $\mathcal{K}$ has countable coproducts. Then so too does $\mathcal{K}^{\underline{Z}}$, and if X

(i) preserves countable coproducts; or

(ii) has a right adjoint

then so does the functor $\overline{X}: \mathcal{K}^{\underline{Z}} \longrightarrow \mathcal{K}^{\underline{Z}}$ defined by

$$(Q\overline{X})_k = Q_{k-1}X$$

and

$$(f\overline{X})_k = f_{k-1}X: Q_{k-1}X \longrightarrow Q'_{k-1}X .$$

Proof: Given $\mathcal{K}^{\underline{Z}}$-objects Q^n, one for each $n \in \underline{N}$ [with $(Q^n)_k = Q^n_k$], we define

$$\left(\coprod_n Q^n\right)_k = \coprod_n Q^n_k \qquad \text{for each } k \in \underline{Z}$$

and it is easy to check that, with the obvious injections, this is a coproduct in $\mathcal{K}^{\underline{Z}}$.

Now, to say that $\overline{X}$ has a right adjoint means we can solve

$$\frac{A\overline{X} \longrightarrow B}{A \longrightarrow B\overline{X}^{\bullet}}$$

for suitable $B\overline{X}^{\bullet}$. But note that we have the correspondence

$$\frac{A_{k-1}X \longrightarrow B_k}{A_{k-1} \longrightarrow B_k X^{\bullet}}$$

so that we define

$$(B\overline{X}^{\bullet})_k = B_{k+1}X^{\bullet} ,$$

and a straightforward computation shows that $\overline{X}^{\bullet}$ is indeed the right adjoint of X. Note that taking the adjoint 'reverses' the direction of 'time'.

We omit the straightforward verification that if $\mathcal{K}$ has, and X preserves, countable coproducts, then $\overline{X}$ too preserves countable coproducts. □

Now an element of $Dyn(\overline{X})$ is just a $\mathcal{K}^{\mathbb{Z}}$-morphism

$$\delta: Q\overline{X} \longrightarrow Q$$

defined by $\delta_k : Q_{k-1}X \longrightarrow Q_k$ for each k --i.e. a time-varying X-dynamics!

10 COROLLARY: If $X: \mathcal{K} \longrightarrow \mathcal{K}$ is an adjoint process, then the time-varying X-process $\overline{X}$ is an adjoint process, with

$$(\overline{X})^{@} = \overline{X}^{*} .$$

Thus

$$[Q\overline{X}^{*}]_k = [\coprod_{n\geq 0} QX^n]_k = \coprod_{n\geq 0} Q_{k-n}X^n . \qquad (3)$$ □

11 EXAMPLE: If $X = id_{\underline{Vect}}$, then $\overline{X} \neq id_{\underline{Vect}^{\mathbb{Z}}}$. In this case (3) yields the weak direct sum

$$[Q\overline{X}^{*}]_k = \coprod_{n\geq 0} Q_{k-n} . \qquad (4)$$

If $X = -\times X_o : \underline{Set} \longrightarrow \underline{Set}$, then

$$[Q\overline{X}^{*}]_k = \coprod_{n\geq 0} Q_{k-n} \times X_o^n ,$$

so that a state of the free dynamics at time k records the q in Q_{k-n} in which the machine 'started', and the string w of X^n_o of inputs received since then.

With this, the theory of time-varying X-dynamics for an adjoint process $X: \mathcal{K} \longrightarrow \mathcal{K}$ reduces to the theory of the adjoint process $\overline{X}: \mathcal{K}^{\underline{Z}} \longrightarrow \mathcal{K}^{\underline{Z}}$. Thus for the theory of reachability, observability, realization and duality of such systems, the reader may turn to [3]. [Bear in mind that if $(\mathcal{E},\mathcal{M})$ is an image factorization system for $\mathcal{K}$, then $(\overline{\mathcal{E}},\overline{\mathcal{M}})$ is an image factorization system for $\mathcal{K}^{\underline{Z}}$, where $\overline{\mathcal{E}} = \{e \mid \text{each } e_k \text{ is in } \mathcal{E}\}$ and $\overline{\mathcal{M}} = \{m \mid \text{each } m_k \text{ is in } \mathcal{M}\}$.]

REFERENCES

1. M. A. Arbib and E. G. Manes: Machines in a Category: An Expository Introduction, <u>SIAM Review</u>, 16, 2 (1974).

2. M. A. Arbib and E. G. Manes: Foundations of System Theory: Decomposable Systems, <u>Automatica</u> (May 1974).

3. M.A. Arbib and E. G. Manes: Adjoint Machines, State-Behavior Machines, and Duality, Technical Report 73B-1, Computer and Information Science, University of Massachusetts at Amherst (January 8, 1973).

4. R. E. Kalman: Algebraic Theory of Linear Systems, in '<u>Topics in Mathematical System Theory</u>' (R. E. Kalman, P. L. Falb, and M. A. Arbib) McGraw-Hill (1969).

5. S. Mac Lane: <u>Categories for the Working Mathematician</u>, Springer-Verlag (1972).

ADDRESSED MACHINES AND DUALITY

E.S. Bainbridge
Mathematics

University of Ottawa
Ottawa, Ont., K1N6N5, Canada

Addressed machines model systems such as a Turing machine tape which have not only a state transition structure but also a data structure. These two structures are dual. A minimal realization theory and duality principle are described for such machines. The minimal realization theory specializes to that for sequential machines [1], algebra automata [4], and linear systems [6]. The duality specializes to the Kalman duality [6] for linear systems and the Arbib-Zeiger duality [1] for sequential machines. Two examples serve to illustrate the underlying ideas.

1. Turing machine tape: Suppose the squares of a Turing machine tape are realized by registers with integer labels representing positions relative to the scanner. A right move of the scanner amounts to the transfer of the contents of each register to its leftmost neighbour. For each letter a_i of the tape alphabet $A = \{a_0,\dots,a_n\}$ let there be a special off-tape square whose contents is always a_i. The information flow among these squares is represented by a diagram

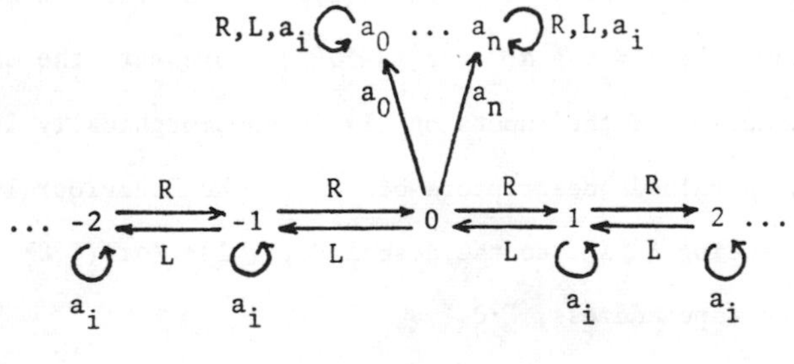

in which there is an arrow labelled x from square t to the square which now contains the symbol which square t will contain after input x. Inputs $x \in X = \{R,L\} + A$ are either scanner moves R,L or print instructions $a_i \in A$. For example, for (print) a_i, all squares except 0 retain their contents; the future contents of square 0 is now in square a_i.

A tape configuration q assigns to each square $t \in T = Z + A$ (where Z is the set of integers) a letter $q(t) \in A$ such that for all squares $a_i \in A$ $q(a_i) = a_i$; and for almost all $n \in Z$, $q(n) = a_0$ (blank). The set Q of tape configurations is a subset of the set A^T of functions from T to A. Transitions between tape configurations induce a right action of the free monoid X^* of input sequences, say $(q,\xi) \mapsto q\cdot\xi$ for $\xi \in X^*$. The diagram above defines a left action of X^* on T, say $(\xi,t) \mapsto \xi\cdot t$ for $\xi \in X^*$. These actions satisfy $(q\cdot\xi)(t) = q(\xi\cdot t)$. For any set S, write $S^\# = A^S$. If a monoid M acts on S from the left, we can define a right action of M on $S^\#$ by $(\sigma\cdot m)(s) = \sigma(m\cdot s)$ for $\sigma \in S^\#$. Thus the inclusion $Q \subseteq A^T$ is a homomorphism of right actions $Q \to T^\#$.

If the tape is provided with a family of initial states $\alpha : I \to Q$ (e.g. a coding of data I) and the output is the contents of the scanned square, $q \mapsto q(0)$, the tape unit is a sequential machine. Note that if 1 is a singleton set and $0 : 1 \to T$ selects $0 \in T$, the output map is $0^\# : T^\# \to 1^\# = A$, $0^\#(q) = q(0)$. The tape unit is represented by the diagram $I \xrightarrow{\alpha} Q \to T^\# \xrightarrow{0^\#} 1^\#$, where the middle map is a homomorphism of right X^* actions. The behaviour of this sequential machine is the function $I \times X^* \to A$ assigning to every datum $i \in I$ and input string $\xi \in X^*$ the contents of the scanned square after the input, $[(\alpha i)\cdot\xi](0)$. The dual diagram $1 \xrightarrow{0} T \to Q^\# \xrightarrow{\alpha^\#} I^\#$ where $t \mapsto (q \mapsto q(t)) : T \to Q^\#$ is a homomorphism of left actions, and $\alpha^\#(\gamma) = \gamma \circ \alpha$ for $\gamma \in Q^\#$, represents the data structure. The left action of the inputs on T is homomorphically interpreted as an action on A-valued descriptors of Q. The behaviour is the function $X^* \to I^\#$ sending $\xi \in X^*$ to the descriptor $i \mapsto (\alpha i)(\xi\cdot 0)$ associated via $\alpha^\#$ with the tape address $\xi\cdot 0$.

2. Algebra Automata: As an example, let A be a finite algebra with a unary operation ν and a binary operation π . Let $A \to 2^L$ be a coding of elements of A as binary L-tuples. Let the operations be defined relative to this coding by equations $(\nu a)_\ell = F_\ell(a_1,\ldots,a_L), (a\pi b)_\ell = G_\ell(a_1,\ldots,a_L, b_1,\ldots,b_L)$, where F_ℓ, G_ℓ $\ell = 1,\ldots,L$ are Boolean expressions in L , 2L variables. For any set S , write $<S>$ for the free Boolean algebra on generators S . Define Boolean homomorphisms $\underline{\nu} : <L> \to <L>$, $\underline{\pi} : <L> \to <L+L>$ as the unique homomorphic extensions of the functions $\ell \mapsto F_\ell$, $\ell \mapsto G_\ell$ respectively. These define a certain coalgebra on $<L>$ in the category of Boolean algebras. For any Boolean algebra B , write $B^\# = \mathrm{Bool}\,[B,2]$ for the set of 2-valued Boolean homomorphisms from B . Then $<L>^\# = 2^L$ becomes an algebra of the same type as A , with operations $(\bar{\nu}\lambda)_\ell = (\underline{\nu}\,\ell)(\lambda_1,\ldots,\lambda_L)$, $(\lambda\bar{\pi}\mu)_\ell = (\underline{\pi}\ell)(\lambda_1,\ldots,\lambda_L, \mu_L,\ldots, \mu_L)$ for $\lambda,\mu \in 2^L$. Thus the inclusion $A \subseteq 2^L$ is a homomorphism of algebras $A \to <L>^\#$, representing the realization of the operations of A on a switching net with junctions L.

3. Definition: For any category A and any functor $u:E \to X$ between small categories, composition with u defines a functor $\Lambda u : A^X \to A^E$, $\Lambda u\,\Phi = \Phi \circ u$ for $\Phi \in A^X$. A left adjoint for Λu is called a left Kan extension [8] of u and is denoted $\Sigma u : A^E \to A^X$. A right adjoint for Λu is called a right Kan extension [8] and is denoted $\Pi u : A^E \to A^X$.

4. Definition Let $(\ )^\# : B \to A$ and $(\ )_\# : A \to B$ be a pair of contravariant functors adjoint on the right. For any functor $\Psi: X \to B$, X a small category, define $\Psi^\# : X^{op} \to A$ by $\Psi^\# x = (\Psi x)^\#$. Likewise, for $\Phi : X \to A$, define $\Phi_\# : X^{op} \to B$ by $\Phi_\# x = (\Phi x)_\#$. If $\psi : \Psi \to \Phi_\#$, define $\psi^* : \Phi \to \psi^\#$ by taking $\psi^* x : \Phi x \to (\Psi x)^\#$ to correspond via the adjunction to $\psi x : \Psi x \to (\Phi x)_\#$. For $\phi : \Phi \to \Psi^\#$, define $\phi_* : \Psi \to \Phi_\#$ similarly.

5. Proposition: Let $u : E \to X$, and let $u^{op} : E^{op} \to X^{op}$ be u reinterpreted as indicated. Let $A, B, {}^\#, {}_\#$ be as in 4, and suppose $\Psi \in B^{X^{op}}$. Then $(\Lambda u^{op}\,\psi)^\# = \Lambda u\,\Psi^\#$, and if $\psi : \Psi \to \Phi_\#$ then $(\Lambda u^{op}\,\psi)^* = \Lambda u\,\psi^*$. Also for $J : E^{op} \to B$, there is a natural isomorphism $(\Sigma u^{op}\,J)^\# \cong \Pi u\,J^\#$ if

either side exists; and various coherence equations relating the adjunctions and this isomorphism hold. Similar relations hold for $(\ )_\#$, $(\ )_*$.

6. Definition: With $u : E \to X$, $\mathcal{A}$, $\mathcal{B}$, $^\#$, $_\#$ as in 5, (see 10 for examples) a machine with input scheme u is a diagram in $\mathcal{A}^E$ of the form $I \xrightarrow{\alpha} \Lambda u\ \Phi \xrightarrow{\beta} J^\#$ for some $\Phi \in \mathcal{A}^X$, $I \in \mathcal{A}^E$, $J \in \mathcal{B}^{E^{op}}$. The behaviour $\underline{B}(M)$ of M is the map in $\mathcal{A}^X$ $\Sigma u\ I \xrightarrow{\overline{\alpha}} \Phi \xrightarrow{\overline{\beta}} \Pi u\ J^\#$ where $\overline{\alpha}$, $\overline{\beta}$ correspond to α, β by the respective Kan adjunctions. An addressed machine K with input scheme u is a diagram in $\mathcal{A}^E$ of the form $I \xrightarrow{\alpha} \Lambda u\ \Phi \xrightarrow{\Lambda u\theta} \Lambda u\ \Psi^\# \xrightarrow{\beta^\#} J^\#$ for some $\Phi \in \mathcal{A}^X$, $\Psi \in \mathcal{B}^{X^{op}}$, $\theta : \Phi \to \Psi^\#$, $I \in \mathcal{A}^E$, $J \in \mathcal{B}^{E^{op}}$, $\beta : J \to \Lambda u^{op}\ \Psi$. The abstract part $\underline{M}(K)$ of K is the machine $I \xrightarrow{\alpha} \Lambda u\ \Phi \xrightarrow{\beta^\# \circ \Lambda u\ \theta} J^\#$. The behaviour of K is $\underline{B}(K) = \underline{B}(\underline{M}(K))$. The dual K^* of K is the addressed machine $J \xrightarrow{\beta} \Lambda u^{op}\ \Psi \xrightarrow{\Lambda u^{op}\ \theta_*} \Lambda u^{op}\ \Phi_\# \xrightarrow{\alpha_\#} I_\#$ with input scheme $u^{op} : E^{op} \to X^{op}$. The data structure of K is the machine $\underline{D}(K) = \underline{M}(K^*)$. Every machine M is the abstract part of the canonically addressed machine $\underline{K}(M)$ $I \xrightarrow{\alpha} \Lambda u\ \Phi \xrightarrow{\Lambda u\ \varepsilon_\Phi} \Lambda u\ {\Phi_\#}^\# \xrightarrow{{\beta_*}^\#} J^\#$ where $\varepsilon_\Phi : \Phi \to {\Phi_\#}^\#$ is a unit of the adjunction.

7. Theorem: (Minimal Realization) Suppose that $\mathcal{A}$ has epi-mono factorizations satisfying a diagonal fill-in theorem [5] (call these strong factorizations). Then every behaviour $B : \Sigma u\ I \to \Pi u\ J^\#$ is realized by a machine $\underline{M}(B)$ of the form $I \xrightarrow{\underline{\phi}} \Lambda u\ \tilde{B} \xrightarrow{\underline{\psi}} J^\#$ where $\Sigma u\ I \xrightarrow{\phi} \tilde{B} \xrightarrow{\psi} \Pi u\ J^\#$ is a strong factorization of B and $\underline{\phi}$, $\underline{\psi}$ correspond to ϕ, ψ via the Kan adjunctions. Moreover, for any other machine M with $\underline{B}(M) = B$ we have a diagram

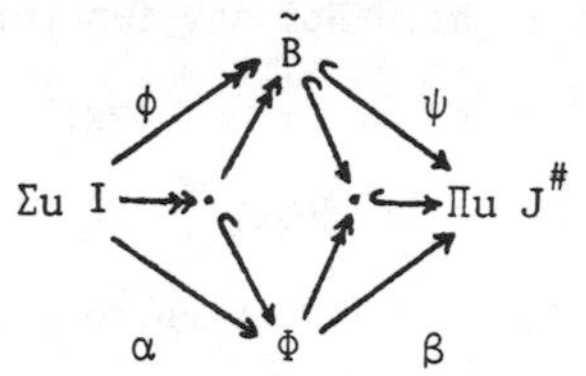

expressing $\underline{M}(B)$ as a quotient of a submachine of M, and as a submachine of a quotient of M.

8. Proposition: (Bistrong factorizations) Let $\mathcal{A}$, $\mathcal{B}$, $\#$, $_\#$ be as in 4, and let X be a small category. If $\mathcal{A}$ and $\mathcal{B}$ have strong factorizations which

are suitably related to ${}^{\#}$, ${}_{\#}$ then every map $\phi : \Phi \to \Psi^{\#}$ in A^{X} has a factorization $\Phi \xrightarrow{\overline{\phi}} \overline{\Phi} \xrightarrow{\varepsilon} \underline{\Psi}^{\#} \xrightarrow{\underline{\phi}^{\#}} \Psi^{\#}$ such that both $\phi = (\underline{\phi}^{\#}\, \varepsilon) \circ \overline{\phi}$ and $\phi_{*} = (\overline{\phi}_{\#}\, \varepsilon_{*}) \circ \underline{\phi}$ are strong factorizations.

9. <u>Theorem</u>: (Biminimal Realization) with data as in 6, 8, every behaviour $B : \Sigma u\, I \to \Pi u\, J^{\#}$ is realized by an addressed machine $\underline{K}(B)$ of the form

$$I \xrightarrow{\phi} \Lambda u\, \tilde{B} \xrightarrow{\Lambda u\, \varepsilon} \Lambda u\, \underset{\sim}{B}^{\#} \xrightarrow{\underline{\psi}^{\#}} J^{\#}$$

where $B = \psi^{\#} \circ \varepsilon \circ \phi$ is the bistrong factorization of 7. Moreover, $\underline{M}(\underline{K}(B)) = \underline{M}(B)$ minimally realizes B, and $\underline{D}(\underline{K}(B)) = M(B_{*})$ minimally realizes B_{*}.

<u>Examples</u>: Sequential machines: Take $A = B =$ Sets, ${}^{\#} = {}_{\#} = Y^{(\ )}$ for some set Y, and $u : \underline{1} \to X^{op}$ where X is the one object category corresponding to the input monoid X and $\underline{1}$ is the trivial one object category. Functors $\Phi : X^{op} \to$ Sets are right actions of X, Λu is the forgetful functor from right actions to sets, machines $1 \xrightarrow{\alpha} \Lambda u\, \Phi \xrightarrow{\beta} 1^{\#} = Y$ are sequential machines with input monoid X, initial state α, output function β with values in Y. The Arbib-Zeiger duality [1] comes from the canonical addressing in that for any M, $\underline{M}(\underline{K}(M)^{*}) = M^{\dagger}$, the Arbib-Zeiger adjoint machine. The minimal realization theory gives the Nerode realization [1] and the data structure of the biminimal realization of $B : X \to Y$ is the machine associated with the <u>left</u> congruence induced by B. The addressing of right (Nerode) congruence classes $\tilde{\xi}_1$ by left congruence classes $\underset{\sim}{\xi}_2$ is specified by the function $(\tilde{\xi}_1, \underset{\sim}{\xi}_2) \mapsto B(\xi_1 \xi_2) \in Y$. The Turing machine tape described in 1 is an example of an addressed sequential machine, as are pushdown stacks and other memory devices.

Linear Systems: The above general theory can accomodate linear systems, but a better treatment comes from a generalization to categories with structured morphism sets, for example additive categories, using Kan extensions for enriched categories [3]. Take $A = B =$ Vect (the (additive) category of finite dimensional vector spaces over a field F), ${}^{\#} = {}_{\#} = (\ )^{*} =$ vector space dual, and $u : F \to F[z]$ the only additive functor where F and its polynomial ring $F[z]$ are interpreted as one object additive categories.

An additive functor $X = F[z] \to \text{Vect}$ is a $F[z]$-module. Λu is the forgetful functor from $F[z]$-modules to vector spaces. A machine $U \to \Lambda u\, X \to (Y^*)^*$ is a linear system. The Kalman duality [6] comes from the canonical addressing in that for any M, $\underline{K}(M)^* = \underline{K}(M^*)$ where M^* is the Kalman dual of M. The minimal realization theory specializes to the usual theory [6] and the data structure of the biminimal realization is the Kalman dual of the minimal realization.

Algebra Automata: Take A = Sets, B = Bool (the category of Boolean algebras), and $B^{\#}$ = Bool $[B,2]$, $S_{\#} = 2^S$. Let T be an algebraic theory category in the sense of Lawvere [7] and let $u : T_0 \to T$ be the embedding of the trivial Lawvere theory. An algebra of type T is a product preserving functor $A : T \to \text{Sets}$. Λu is the forgetful functor from T-algebras to sets. A machine $I \xrightarrow{\alpha} \Lambda u\, A \to (Y_{\#})^{\#} = Y$ where α factors through the unique homomorphism $T(\emptyset) \to A$ from the initial T-algebra $T(\emptyset)$ is an algebra automaton. The minimal realization theory specializes to the usual theory [4] (details may be found in [2]). The data structure of the biminimal realization is the coalgebra of A-valued descriptors of the minimal realization.

REFERENCES

[1] ARBIB, M.A. and ZEIGER, H.P. "An Automata-Theoretic Approach to Linear Systems" IFAC Symposium, Sydney Australia, August 1968.

[2] BAINBRIDGE, E.S. "A Unified Minimal Realization Theory, with Duality". Dissertation, Department of Computer & Communication Sciences, University of Michigan, 1972.

[3] DUBUC, E.J. Kan Extensions in Enriched Category Theory, Lecture Notes in Mathematics #145, Springer-Verlag, New York 1970.

[4] GIVE'ON, Y. and ARBIB, M.A. "Algebra Automata II: The Categorical Framework for Dynamic Analysis" Information and Control, 12, pp. 346-310, 1968.

[5] GRILLET, P.A. "Regular Categories", in Exact Categories and Categories of Sheaves, Lecture Notes in Mathematics #236, Springer-Verlag, New York 1971.

[6] KALMAN, R.E. "Algebraic Theory of Linear Systems", in Topics in Mathematical System Theory. edited by R.E. Kalman, M.A. Arbib, and P.L. Falb. McGraw-Hill, 1969.

[7] LAWVERE, F.W. "Functorial Semantics of Algebra Theories", Proc. Nat. Acad. Sci., U.S.A., 50, pp. 869 - 872, 1963.

[8] MACLANE, S. Categories for the Working Mathematician, Springer-Verlag, New York, 1971.

FACTORIZATION OF SCOTT-STYLE AUTOMATA

John L. Baker
Computer Science

University of British Columbia
Vancouver, British Columbia, Canada V6T 1W5

Scott (1967) suggests developing automata theory along the following lines: There should be only one machine of each species (e.g. turing machine, n-dimensional bug, RAM), each such machine being programmable. The languages and sets of computations whose specification is the objective of automata theory should each be determined by a *program* for some such machine. For example, in this scheme of things the statement

(1) L is a context-free language over the alphabet A

is shown to be equivalent to

(2) There is a program Π for *the* pushdown automaton with alphabet A such that L is the set of outputs of computations under control of Π

In illustrating the application of his definitional suggestions to familiar portions of automata theory, Scott finds it necessary to introduce the notion of a *machine with standard input-output*. Such a machine is one the set of whose memory configurations is a Cartesian product of input, output, and working components. The execution of each step of a program for such a machine influences, and is dependent on, at most one of the three components mentioned. Interaction among these components is achieved only by way of control flow in the program. Furthermore, Scott uses an arbitrary fixed alphabet to specify the exact form of the input and output components, and of the program steps involving them.

Eliminating the specialization to input-output mentioned in the last sentence, one obtains a notion of *product of machines*, applicable not only to the regularizing of input-output details of computations, but also to the presentation of complex memory structures (multiple tapes or push-down stores, *etc.*) which are important in at least some parts of automata theory. A formal definition of product of machines appears below.

The advantage claimed in Scott (1967) for his approach is *uniformity:* Many definitions relating to machines and their computations can be given once for all and theorems expressing the nature of computation under finite control *per se* can be stated and proved in the highest generality.

The significance of this advantage depends, of course, on the *viability* of the approach: Does it permit the perspicuous expression of the fundamental concepts and results of automata theory as now understood, and does it suggest new ideas?

Scott (1967) makes a good case for the expressive power of this approach, presenting several familiar machines and an outline of an argument for (1) ≡ (2). The following new idea suggested by this approach is also exposed there: The notion of partial function computable by machine M logically precedes any notion of set recognizable by M. It is reasonable, then, to consider as sets of interest the ranges, domains, and inverse images of points with respect to functions computable by a machine M. (These turn out to be the familiar (possibly non-deterministically-) recognizable sets for M.) To put it another way, the new idea is to take seriously the *direct* generalization of the definitions of partial recursive function, recursive and recursively enumerable set to restricted notions of computability.

Over the past two years, I have developed Scott's suggestions in considerable detail, and used my development in teaching. The reaction of colleagues and students has been favorable. (It should be mentioned that, because of the uniformity of this approach, it is possible to specify a single "programming language" for all automata, the use of which brings the intuition of the computer scientist to bear directly on automata-theoretic questions. Conversely, the use of programming techniques in a setting where rigor is habitual has a favorable influence on ordinary programming.) As an example of my development, I offer the characterization of full AFL given below, which I venture to say is radically clearer, both in statement and proof, than that of Ginsburg and Greibach (1969).

It is in search of further new ideas in automata theory that I submit this abstract to the Symposium. I believe that the approach presented here is clear and orderly, and I am eager to know how it can be melded with the ideas of categorical algebra to produce a structure of even greater clarity.

The rest of this abstract will comprise a telegraphic statement of definitions and the full AFL result mentioned above.

NOTATIONS

$1 = \{0\}$.

λ denotes the empty string, the only string of length 0.

if $x = a_1 a_2 \ldots a_{n-1} a_n$ is a string of length n,

then $(x] = a_2 \ldots a_{n-1} a_n$. $(\lambda] = \lambda$

A *function* f (ordinarily called a *partial function*) is determined by a *source set* S, *a target set* T, and a univocal relation $\mapsto$ in $S \times T$ (the *graph* of f). We write $f\colon S \to T : x \mapsto f(x)$.

dom $(f) = \{x \mid \exists_y x \mapsto y\}$. ran $(f) = \{y \mid \exists_x x \mapsto y\}$.

F and P are sets, F, P, and $\{In, Out\}$ pairwise disjoint. (F, P are *uninterpreted functions* and *predicates*, respectively.)

DEFINITIONS

1. An F - P - program Π is a finite set of strings called *instructions*, satisfying

   (a) Exactly one instruction of Π is of the form

   (i) $\uparrow Q_1$ ("start at Q_1")

   (b) All other instructions of Π are of one of the forms

   (ii) $Q_0 : F; Q_1$ ($F \in \mathsf{F}$) ("perform F, then continue at Q_1")

   (iii) $Q_0 : P \Rightarrow Q_1; Q_2$ ($P \in \mathsf{P}$) ("if P, then continue at Q_1 otherwise at Q_2")

   (iv) $Q_0 : \downarrow$ ("Halt")

   (c) The Q_i are elements of a set $\mathcal{Q}_\Pi$, disjoint from F and P, the set of *labels occuring* in Π. Each element of $\mathcal{Q}_\Pi$ must occur exactly once in Π in the position indicated above as Q_0, where it is said to be the *label of* the instruction in which it so occurs.

2. An F - P - machine M is a function defined on $\mathsf{F} \cup \mathsf{P} \cup \{In, Out\}$ such that there are sets X (input), Y (output), and M (memory), and functions

$M_{In} : X \to M$ (the input function of M)

$M_{Out} : M \to Y$ (the output function of M)

$\forall_{F \in \mathsf{F}}$ $M_F : M \to M$ (the M-interpretation of F)

$\forall_{P \in \mathsf{P}}$ $M_P : M \to \{true, false\}$ (the M-interpretation of P).

3. If M is an F - P - machine with memory set M, Π an F - P - program, the binary relation $\Rightarrow_{M,\Pi}$ is defined on $L_\Pi \times M$ by:

$$<Q_0,m> \Rightarrow_{M,\Pi} <Q_1,M_F(m)> \text{ if } Q_0 : F;\ Q_1 \in \Pi$$
$$<Q_0,m> \Rightarrow_{M,\Pi} <Q_1,m> \text{ if } Q_0 : P \Rightarrow Q_1;\ Q_2 \in \Pi \text{ and } M_P\ (m) = \mathit{true}$$
$$<Q_0,m> \Rightarrow_{M,\Pi} <Q_2,m> \text{ if } Q_0 : P \Rightarrow Q_1;\ Q_2 \in \Pi \text{ and } M_P\ (m) = \mathit{false}$$
$$<Q_0,m> \not\Rightarrow_{M,\Pi} <Q_1,m> \text{ otherwise.}$$

$\overset{*}{\Rightarrow}_{M,\Pi}$ is defined to be the reflexive-transitive closure of $\Rightarrow_{M,\Pi}$

4. If M is an F - P -machine with input set X, output set Y; and Π is an F - P - program, then the *function computed by* M *under* Π is

$$M_\Pi : X \to Y : x \mapsto M_{Out}(m) \text{ , where}$$
$$\uparrow Q_0 \in \Pi,\ <Q_0,M_{In}(x)> \overset{*}{\Rightarrow}_{M,\Pi} <Q_1,m>, \text{ and } Q_1 : \downarrow \in \Pi.$$

5. If, for $i = 1,2,\ldots,n, M_i$ is an F_i - P_i - machine with input, output, memory sets X_i, Y_i, M_i and the F_i, P_i are all pairwise disjoint, then the product

$$M_1 \times M_2 \ldots \times M_n$$

is a $(\cup_{i=1}^{n} F_i)$ - $(\cup_{i=1}^{n} P_i)$ - machine M defined thus:

Let: $X = X_{i_1} \ldots \times X_{i_k}$, where $i_1 \ldots <i_k$ and

$$\{i_s \mid 1 \leq s \leq k\} = \{i \mid 1 \leq i \leq n \text{ and card } (X_i) \neq 1\}$$

$Y = Y_{j_1} \ldots \times Y_{j_l}$, where $j_1 \ldots <j_l$ and

$$\{j_s \mid 1 \leq s \leq l\} = \{i \mid 1 \leq i \leq n \text{ and card } (Y_i) \neq 1\}$$

$M = M_1 \times M_2 \ldots\ M_n$

Define: $M_{In} : X \to M : <x_{i_1} \ldots, x_{i_k}> \mapsto <M_{1,In}(x_1), M_{2,In}(x_2) \ldots, M_{n,In}(x_n)>$

where, if card $(X_i) = 1$, $X_i = \{x_i\}$.

$M_{Out}: M \to Y : <m_1,m_2 \ldots,m_n> \mapsto <M_{j_1,Out}(m_{j_1}) \ldots, M_{j_l,Out}(m_{j_l})>$

provided $\forall_{i=1}^{n} M_{i,Out}(m_i)$ is defined.

$M_F : M \to M : <m \ldots, m_i \ldots, m_n> \mapsto <m \ldots, M_{i,F}(m_i) \ldots, m_n>$

$(F \in F_i,\ 1 \leq i \leq n)$

$M_P : M \to \{\mathit{true},\mathit{false}\} : <m_1 \ldots, m_i \ldots, m_n> \mapsto M_{i,P}(m_i)$

$(P \in P_i,\ 1 \leq i \leq n)$

6. If A is a finite set, $Inl^{(A)}$, the *one-way input machine over A*, is an $\{\leftarrow I\}$-$\{I{=}a \mid a\varepsilon A\}$-machine satisfying

$$Inl^{(A)}_{In} \quad : A^* \to A^* : x \mapsto x$$

$$Inl^{(A)}_{Out} \quad : A^* \to 1 \quad : x \mapsto 0$$

$$Inl^{(A)}_{\leftarrow I} \quad : x \to (x]$$

$$Inl^{(A)}_{I=a}(x) \equiv x = a(x] \qquad (a\varepsilon A)$$

7. Aux, the *auxiliary input machine*, is an $\{\leftarrow A\}$-$\{A\}$-machine satisfyiing

$$Aux_{In} : \{true, false\}^* \to \{true, false\}^* : x \mapsto x$$

$$Aux_{Out}: \{true\ false\}^* \to 1 : x \mapsto 0$$

$$Aux_{\leftarrow A} : x \mapsto (x]$$

$$Aux_{A} \quad (x) \equiv x = true(x].$$

8. If A is a finite set, $Out^{(A)}$, the *output machine over A*, is an $\{0{\leftarrow}a \mid a\varepsilon A\}$-$\emptyset$-machine satisfying

$$Out^{(A)}_{In} \quad : 1 \to A^* : 0 \mapsto \lambda$$

$$Out^{A}_{Out} \quad : A^* \to A^* : x \mapsto x$$

$$Out_{0\leftarrow a} \quad : x \mapsto xa \qquad (a \ \varepsilon \ A)$$

(SAMPLE) RESULT

9. The following are equivalent:

(a) $\langle\Sigma, L\rangle$ is a full AFL. (That is, L is a family of languages, each over a finite subset of Σ, including a non-empty language and closed under union, * (closure), inverse homomorphism, homomorphism, and intersection with regular set.)

(b) There is an F-P-machine M satisfying

(i) M has input set = output set = 1 and

(ii) (Reset condition) There is a set of insturctions Γ such that $\langle Q_0, m\rangle \overset{*}{\Rightarrow}_{M,\Gamma} \langle Q_1, M_{In}(0)\rangle$ for all elements m of the memory set of M, where Q_0, $Q_1 \varepsilon\ Q_\Gamma$

such that

$L = \{ran(Aux \times M \times Out^{(A)}_{\Pi}) \mid A$ is a finite subset of Σ and Π is an $F\cup\{\leftarrow A\}\cup\{0{\leftarrow}a \mid a\varepsilon A\}$-$P\cup\{A\}$ program.$\}$

Proof: (a) ⇒ (b) : Define

$M_{In} : 1 \to \Sigma^* : 0 \mapsto \lambda$

$M_{Out} : \Sigma^* \to 1 : \lambda \mapsto 0$, undefined on Σ^+

$M_{Z\leftarrow a} : x \mapsto xa \quad (a\varepsilon\Sigma)$

$M_{Z!L} : x \mapsto \lambda$ **if** $x \varepsilon L$, undefined off L ($L \varepsilon \mathcal{L}\cup\{\Sigma^*\}$)

M satisfies (i) by definition, (ii) By $\Gamma=\{Q_0:Z!\Sigma^*;Q_1\}$. If $L \varepsilon \mathcal{L}$, $L\subset A^*$, then $L = \mathrm{ran}(Aux\mathrm{x}M\mathrm{x}Out^{(A)}\Pi)$, where Π interprets the input as a code for an element x of A^*, copies x to the memory of M, then executes $Z!L$.

Conversely, let Π and A be given. Let $L_1, L_2 \ldots, L_n$ be the languages mentioned in $Z!L$ instructions of Π. (Without loss of generality, suppose Σ^* is not among these L_i.) Obtain Π' from Π by replacing $Z\leftarrow a$ by $0\leftarrow\bar{a}$, $Z!L_i$ by $0\leftarrow\$_i$. Define $R= \mathrm{ran}(Aux\mathrm{x}Out^{(A')}{}_{\Pi'})$, a regular set. Then

$$\mathrm{ran}(Aux\mathrm{x}M\mathrm{x}Out^{(A)}{}_{\Pi}) = f(g^{-1}((\cup_{i=1}^{n}(L;\{\$_i\}))^*)\cap R),$$

where $f : \bar{a} \mapsto a$ if $a \varepsilon A$, $a \mapsto \lambda$ otherwise; $g : \bar{a} \mapsto \lambda$ **if** $a \varepsilon A$ $a \mapsto a$ otherwise.

(b) ⇒ (a) Closure of $\mathcal{L}$ under $\cup$ and * is conveniently enough shown by programming constructions which analyze the input and pass portions of it to given language generators used as subroutines. The construction for * depends critically on the hypothesized reset instructions.

Closure of $\mathcal{L}$ under the remaining operations is shown using the following two lemmas:

(3) $\mathcal{L} = \{\{x|\exists_y <x,y>\varepsilon \text{ dom } (In1^{(A)}\mathrm{x}Aux\mathrm{x}M_{\Pi})\}|A$ is a finite subset of Σ and Π is an $F\cup\{\leftarrow I,\leftarrow A\}-P\cup\{A\}\cup\{I=a|a\varepsilon A\}$-program$\}$

(4) If $\exists_{\Pi_1} f_1 = (In1^{(A)}\mathrm{x}N_1)_{\Pi_1}$ and $\exists_{\Pi_2} f_2 = (N_2\mathrm{x}Out^{(A)})_{\Pi_2}$, then

$\exists_{\Pi} f_1 \circ f_2 = (N_2\mathrm{x}N_1)_{\Pi}$, providing all machines involved are defined.

Proof of (3) and (4) are again straight forward programming constructions. (3) is rather tedious and technical, a matter of coding two inputs into one, but (4) is rather attractive, a matter of linking Π_2 (producer) and Π_1 (consumer) as coroutines. We couple these lemmas with the observations that homomorphisms $h:A^*\to B^*$ are $In1^{(A)}\mathrm{x}Out^{(B)}$-computable, and that the restriction i_R of the identity function to a regular set $R\subset A^*$ is $In1^{(A)}\mathrm{x}Out^{(A)}$-computable to obtain $h(L)$, $h^{-1}(L)$, and $L\cap R$ as ranges or projections of domains of appropriately computable functions whenever $L\varepsilon\mathcal{L}$.

REFERENCES

Ginsburg, Seymour, and Sheila Greibach (1969). Abstract families of languages. American Mathematical Society, *Memoir* 87, pp. 1:32.

Scott, Dana (1967). Some definitional suggestions for automata theory. *Journal of Computer and System Sciences,* v.1., pp. 187:212.

AN ABSTRACT MACHINE THEORY FOR FORMAL LANGUAGE PARSERS*

David B. Benson
Computer Science Department

Washington State University
Pullman, WA 99163, USA

An abstract machine theory specifies the nature of the machines in terms of the desired properties of the transition, output and initial state selection functions. Developed in appropriate categories, one obtains notions of minimum machines. In this paper our interest is in developing such a theory for formal language parsers, a form of unbounded automata. A typical [2,3,9,10] theory is developed with a nice surprise--minimal machines have the important property of immediate error detection. This theory is not a special case of [2,3,10] and gives the only known classification of all the uniform left-to-right parsers for a given grammar.

(Σ,P) is a rewrite system if Σ is a set and P is a function from some indexing set J to $\Sigma^* x \Sigma^*$. Lower case Greek letters denote strings in Σ^*. λ denotes the null string. For $j \epsilon J$, P(j) is written in the form $\alpha \rightarrow \beta$, and is called a production or rewrite rule. γ derives δ if $\gamma = \mu\alpha\nu$, $\delta = \mu\beta\nu$ and for some $j \epsilon J$, $P(j) = \alpha \rightarrow \beta$. The transitive and reflexive closure of derives is written $\alpha \Rightarrow \beta$ and one says that there is a derivation from α to β.

---

*The paper is to appear in Acta Informicata. Research supported in part by the National Science Foundation Grant GJ-1171.

(Σ,P,ζ) is a pointed rewrite system when (Σ,P) is a rewrite system and $\zeta\varepsilon\Sigma^*$. Think of ζ as the axiom from which one derives. If one distinguishes a subset, E, of Σ as the "terminal" or external letters, then the formal language generated by the grammer $G=(\Sigma,E,P,\zeta)$ is $L(G)=\{\omega\varepsilon E^* \mid \zeta\Rightarrow\omega\}$.

A parsing function for G assigns to each $\omega\varepsilon L(G)$ the set of all derivations from ζ to ω and assigns to each $\omega\notin L(G)$ the empty set, these ideas to be made precise shortly. The notion abstracts the usual ideas of parsing algorithms used in computational linguistics and programming language compilers.

The concept of derivation must be made precise. This is accomplished via syntax categories, to be established by the following definitions.

A category is a list $\underset{\sim}{C}=(O,M,d,c,\circ)$ where O is the set of objects, M the morphisms or arrows, $d:M\to O$, $c:M\to O$ the domain and codomain functions and $\circ$ the composition of morphisms, all subject to the usual axioms [14,15]. A strict monoidal category (sm-category) $(\underset{\sim}{C},\Box,e)$ is a category $\underset{\sim}{C}$ together with an associative bifunctor $\Box$ and an object e which is the identity for $\Box$, so that O becomes a monoid under the action of $\Box$ and so does M, [14].

A free sm-category is generated by a rewrite system (Σ,P) as follows: Let Σ^* be the objects, P be the generators of the morphisms, M, with the bifunctor + as the free monoid operation on Σ^*, and parallel derivations on M. Clearly $\lambda\varepsilon\Sigma^*$ is the identity for the free sm-category $\underset{\sim}{X}(\Sigma,P)$. $\underset{\sim}{X}(\Sigma,P)$ is called the syntax category of (Σ,P) and each morphism is a derivation from its domain to its codomain, [11,6]. Note that for derivations $x:\alpha\to\beta$, $y:\gamma\to\delta$, $x\circ y$ is defined iff $c(x)=d(y)$ while $x+y$ is always defined to be the parallel derivation from $\alpha+\gamma$ to $\beta+\delta$.

With this, we see that a parsing function is just a restriction of the external representation functor (hom functor) $h^{\zeta}:\mathbf{X}\to\mathcal{S}$ with $h^{\zeta}(\omega) = \hom_{\mathbf{X}}(\zeta,\omega)$. [14,15]. We can extend further by considering all possible points, ζ, to arrive at the external representation bifunctor (the hom bifunctor), $h:\mathbf{X}^{op}\times\mathbf{X}\to\mathcal{S}$ where $\mathbf{X}^{op}$ is the opposite category to $\mathbf{X}$.

We consider $h:\mathbf{X}^{op}\times\mathbf{X}\to\mathcal{S}$ to be the natural abstract parsing behavior of an automaton which parses a formal language [1,4,5,7,8,12,13].

In this rarified atmosphere, we next give an appropriate notion of abstract automata.

An abstract parsing machine for $\mathbf{X}=\mathbf{X}(\Sigma,P)$ is $A=(\mathbf{C},\mathbf{X},a,g,i)$ where $\mathbf{C}$ is a small category, a is a right action for $\mathbf{C}$ and $\mathbf{X}$, $g:\mathbf{C}\to\mathcal{S}$ and $i:\mathbf{X}^{op}\to\mathbf{C}$ are functors such that the behavior of A is the external representation bifunctor for $\mathbf{X}$. These notions are now explained.

A right action for $\mathbf{C}$ and $\mathbf{X}$, [14], is a bifunctor $a:\mathbf{C}\times\mathbf{X}\to\mathbf{C}$ making the following diagram commute.

$$\begin{array}{ccc} \mathbf{C}\times\mathbf{X}\times\mathbf{X} & \xrightarrow{\;1_{\mathbf{C}}\times +\;} & \mathbf{C}\times\mathbf{X} \\ \Big\downarrow{\scriptstyle a\times 1_{\mathbf{X}}} & & \Big\downarrow{\scriptstyle a} \\ \mathbf{C}\times\mathbf{X} & \xrightarrow[\;a\;]{} & \mathbf{C} \end{array}$$

That is, $a(c,x_1+x_2) = a(a(c,x_1),x_2)$ for c in $\mathbf{C}$, x_1,x_2 in $\mathbf{X}$.

The behavior of A is the functor $(i\times 1_{\mathbf{X}})ag:\mathbf{X}^{op}\times\mathbf{X}\to\mathcal{S}$, i.e., $g(a(i(x_1^{op}),x_2))$ for x_1^{op} in $\mathbf{X}^{op}$, x_2 in $\mathbf{X}$.

Relating to more ordinary automata, $\mathbf{C}$ is the states, the right action a corresponds to the transition function, g corresponds to the output function, while i selects the initial state.

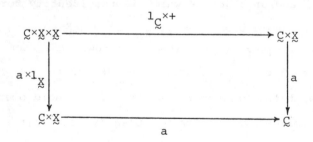

For concrete parsers, various methods are used to concisely represent the objects of $\underset{\sim}{C}$. In [4], Hibbard graphs are used. For context free parsers, trees or pushdown lists are used [1,7,12, 13]. The morphisms of $\underset{\sim}{C}$ are not represented as data. They describe the action of the parser upon discovering the applicability of some production or indeed, a derivation. For <u>shift-reduce</u> parsers [1,12], the right action corresponds to a shift followed by zero or more reductions. The morphisms of $\underset{\sim}{C}$ are then the potential reduction sequences. The output functor g corresponds to selecting the appropriate parses from the data used to control the parser. For example, in [4,8] the control data represents many partial parses. When parsing is complete, the full parses must be selected from this data.

A morphism from $A=(\underset{\sim}{C},\underset{\sim}{X},a,g,i)$ to $A'=(\underset{\sim}{C}',X',a',g',i')$ is a functor $k:\underset{\sim}{C}\to\underset{\sim}{C}'$ together with a strict monoidal functor $j:\underset{\sim}{X}\to\underset{\sim}{X}'$ satisfying the following

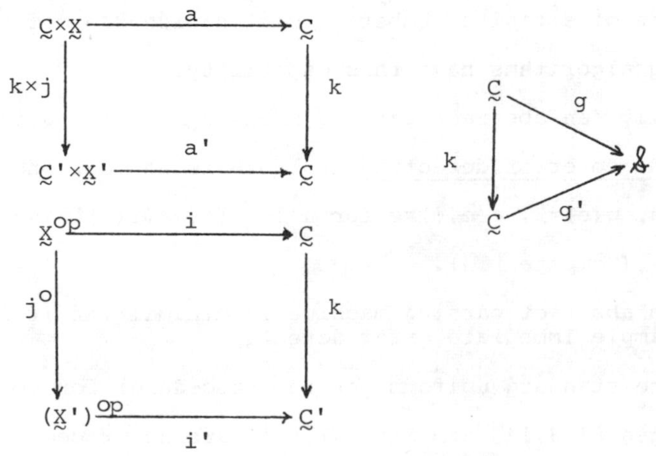

With this let $\underset{\sim}{M}$ denote the category of all abstract parsing machines and $\underset{\sim}{M}_{\underset{\sim}{X}}$ denote the full subcategory of abstract parsing machines with input $\underset{\sim}{X}$.

Our results are stated without proof.

1. The external representation bifunctor, properly construed, is an initial object of $\underset{\sim}{M}_{\underset{\sim}{X}}$.

$(\underset{\sim}{C},\underset{\sim}{X},a,g,i)$ is reachable if $\underset{\sim}{C}$ is the image of the functor $a(i(-),-):\underset{\sim}{X}^{op}\times\underset{\sim}{X}\to\underset{\sim}{C}$.

2. If $(\underset{\sim}{C},\underset{\sim}{X},a,g,i)$ is reachable then the $\underset{\sim}{M}_{\underset{\sim}{X}}$-morphism $(a(i(-),-),1_{\underset{\sim}{X}})$ is an epimorphism, but not conversely due to pathological rewriting systems.

Let $\underset{\sim}{R}_{\underset{\sim}{X}}$ be the full subcategory of $\underset{\sim}{M}_{\underset{\sim}{X}}$ with objects the reachable abstract parsing machines. Define a minimal parser to be a final object in $\underset{\sim}{R}_{\underset{\sim}{X}}$.

3. There is a minimal abstract parsing machine for each $\underset{\sim}{X}$, by a Nerode construction.

We will say that a string θ is ζ-wrong whenever it is not the prefix of any string derivable from ζ, i.e., if for all $\psi\varepsilon\Sigma^*$, $\hom_{\underset{\sim}{X}}(\zeta,\theta+\psi)=\emptyset$.

A parser has immediate error detection ability if whenever a ζ-wrong input is presented, for point ζ, the parser immediately enters one of a finite number of distinguished error states. Good parsing algorithms have this capability.

Formally, an abstract parsing machine $A=(\underset{\sim}{C},\underset{\sim}{X},a,g,i)$ is simple immediate error detecting if there is at most one object $e\varepsilon\underset{\sim}{C}$ such that $g(e)=\emptyset$, $a(e,\psi)=e$ for all $\psi\varepsilon\Sigma^*$, $a(i(\zeta),\theta)=e$ iff θ is ζ-wrong. (Compare [1]).

4. An abstract parsing machine is minimal iff it is simple immediate error detecting.

All the standard uniform parsing procedures for context free languages [1,8,12] are minimal. A parsing procedure for Chomsky Type 0 grammars is initial, [4].

Our methods give control over the number of "error states," but do not give any control over the internal complexity of each object in the state category $\underset{\sim}{C}$. Further research will be required to completely capture the intuitive notions of a minimal parser.

REFERENCES

1. Aho, A.V., and Ullman, J.D.: The Theory of Parsing, Translation and Compiling, Prentice-Hall, 1972.

2. Arbib, M.A., and Manes, E.G.: Machines in a Category, an Expository Introduction, **SIAM Review, to appear.**

3. Arbib, M.A., and Manes, E.S.: Adjoint Machines, State-Behavior Machines and Duality, **Journal of Pure and Applied Algebra, to appear.**

4. Benson, D. B.: The Algebra of Derivations and a Semithue Parser, Proc. 24th Nat. ACM Conf., 1969, 1-9.

5. Benson, D.B.: Syntactic Clues, IEEE Conf. Record 11th Annual Symp. on Switching and Automata Theory, 1970, 133-138.

6. Benson, D.B.: The Basic Algebraic Structures in Categories of Derivations, Comp. Sci. CS-73-004, Washington State University. To appear in Information and Control.

7. DeRemer, F.L.: Simple LR(k) Grammars, CACM 14, 1971, 453-460.

8. Earley, J.: An Efficient Context-Free Parsing Algorithm, CACM 13, 1970, 94-102.

9. Give'on, Y.: A Categorical Review of Algebra Automata and Systems Theories, Symposia Mathematica, Vol. 4, Institute Nazionale di Alta Matematica, Academic Press, 1970.

10. Goguen, J.A.: Minimal Realization of Machines in Closed Categories, Bull. Am. Math. Soc. 78, 1972, 777-783.

11. Hotz, G.: Eindeutigkeit und Mehrdeutigkeit Formaler Sprachen, EIK 2, 1966, 235-247.

12. Knuth, D.E.: On the Translation of Languages from Left to Right, Information and Control 8, 1965, 607-639.

13. McKeeman, W.M., et.al.: A Compiler Generator, Prentice-Hall, 1970.

14. MacLane, S.: Categories for the Working Mathematician, Springer, 1972.

15. Pareigis, B.: Categories and Functors, Academic Press, 1970.

SOME STRUCTURAL PROPERTIES OF AUTOMATA DEFINED ON GROUPS

Roger W. Brockett
Division of Engineering and Applied Physics
Harvard University
Cambridge, Massachusetts 02138 U.S.A.

Alan S. Willsky
Department of Electrical Engineering
Massachusetts Institute of Technology
Cambridge, Massachusetts 02139 U.S.A.

The study of dynamical systems defined on groups has proceeded along two lines, one devoted to the finite group case, the other to continuous groups. In each case there has been a structural theory developed which bears some relationship to the structure theory for finite dimensional linear systems. In this abstract we sketch a theory which captures the essential features of the isomorphism theorems in [1] (theorem 6) and [2] (theorem 10) in a single framework. We also raise some questions relating to further generalizations of these ideas.

Let (X,U,Y) be a triple of groups. By a group homomorphic sequential system on (X,U,Y) we mean a triple of group homomorphisms (a,b,c) whereby $a : X \to X$; $b : U \to X$; $c : X \to Y$ with the dynamical action being described via

$$x(k+1) = b[u(k)]\cdot a[x(k)]; \quad y(k) = c[x(k)] \qquad (1)$$

This work was supported in part by the U.S. Office of Naval Research under the Joint Services Electronics Program by Contract N00014-67-A-0298-0006 (R.W. Brockett) and by NSF under Grant NSF-GK-25781 (A. Willsky).

Added structure leads to a variety of modifications; e.g. one can ask that the groups involved be finite, that they be topological with (a,b,c) continuous, that they be Lie groups, etc. By Theorem 6 of [1] we know that two controllable and observable finite group homomorphic realizations of the same input-output map are isomorphic.

On the other hand, in the study of systems defined on Lie groups, continuous time models of the form

$$\dot{x}(t) = (a + \sum_{i=1}^{m} u_i(t)b_i)x(t); \quad y(t) = c[x(t)] \tag{2}$$

have been studied enough to obtain a structure theory. Here x belongs to a Lie group X, ax and bx define vector fields on the Lie group which are right invariant (i.e. if we translate x to xm by means of right multiplication by an element of X then $\frac{d}{dt}(xm) = a(xm)$, etc.). The controls $u_i(t)$ belong to the reals and $c : X \to Y$ such that c is one to one on the coset space X/C, for some subgroup C. This particular setup covers some cases arising in applications and, like the group homomorphic case, it has a nice mathematical theory associated with it. In particular, by theorem 10 of [2] we know that the controllable and observable realizations of the same input-output map have isomorphic Lie algebras.

The differences between these two cases is perhaps more apparent than the similarities. To facilitate comparison we look at a difference equation, version of (2). Let X be a group and C a subgroup of X. Let U and Y be sets. Consider

$$x(k+1) = b[u(k)]\cdot a\cdot x(k); \quad y(k) = c[x(k)] \tag{3}$$

where $b : U \to X$ and $c : X \to Y$ is one to one on the coset space X/C. We call such a system a <u>group translation sequential system</u> on (U,X,Y). Notice that: i) we have abandoned the idea that U and Y should be groups, ii) since $a\cdot x$ does not (except for the case a = id) define a homomorphism, we do not have the same kind of dynamical action as in the homomorphic case and, iii) since the state at time k contains the starting state as a factor, even if we give Y a group structure we cannot have observability with c a homomorphism (and therefore C normal in X) unless C is trivial.

It is possible to study equation (1) on Lie groups; see [7] for some results on controllability and observability. One problem which arises in this framework is that for Lie groups, automorphisms look a great deal like inner automorphisms, (see Hausner and Schwartz [4] page 180, 181). For inner automorphisms a, the group machine dynamics looks like

$$x(k+1) = b[u(k)]ax(k)a^{-1}; \quad y(k) = c[x(k)] \tag{4}$$

For observability in the sense of [1] we require that $c(a^k x(0)a^{-k})$, $k=0,1,\ldots$ determine $x(0)$. But $c(a^k x(0)c^{-k}) = c(a^k)c(x(0))c(a^{-k})$ so that the system is observable only in the relatively trivial case where c is injective.

These facts necessitate a different point of view -- one suggested by the following theorem. (Compare with [2] theorem 10.)

Theorem 1: Let X and Z be finite groups. Let the two group translation sequential systems

$$x(k+1) = b[u(k)\cdot a\cdot x(k); \quad y = c(x(k)); \quad x(0) = \text{id}.$$

and

$$z(k+1) = g[u(k)]*f*z(k); \; y(k) = h[z(k)]; \; z(0) = \text{id}.$$

define the same input/output map. Suppose that both systems are controllable in that any state can be reached from any other and suppose that both are observable in that any two distinct states give rise to distinct outputs for some input string. Then X and Z are isomorphic as groups and the map

$$\eta : a \to f$$

$$\eta : b(u_i) \to g(u_i)$$

extends to a group isomorphism $\eta : X \to Z$.

Proof: By the controllability hypothesis the products $b[u_1]a$, $b[u_2]a$, ... and $g(u_1)f$, $g(u_2)f$,... generate X and Z respectively. Moreover, if some product of these equal the identity in X they also do in Z since by assumption the identity states in X and Z are equivalent in the snese of Nerode (see proof of theorem 3.) Thus for any relation $\Pi[b(u_i)a] = \text{id}$ there is a corresponding relation $\Pi[g(u_i)f] = \text{id}$. Now describing X and Z by their generators and relations we see that X and Z are isomorphic.

Notice that in [1] the isomorphism theorem is proven by appealing to the fact that the output map is a homomorphism. In [2] the isomorphism theorem is proven in the above way, working with the controllability hypothesis. We will later indicate how this kind of proof can be generalized.

Now to obtain a result which includes the state space isomorphism theorem for finite group homomorphic (actually automorphic) sequential systems and finite group translation systems as special cases, we introduce a new and more general class of machines. Let X be a group. By a <u>holomorphism</u> one means a one-to-one map f of X into X such that

$$f(xy^{-1}z) = f(x)(f(y))^{-1}f(z)$$

The set of all holomorphisms is itself a subgroup of the group of all permutations on X. As such it is generated by the set of left translations and the set of automorphisms (see [3] page 20); thus it contains both as special cases. In fact f is holomorphic if and only if $f(\cdot) = \alpha\cdot\beta(\cdot)$.

<u>Definition</u>: Let U and Y be sets and let X be a group. Let $a : X \to X$ be a holomorphism and let $b : U \to X$ and $c : X \to Y$ subject to the conditions that $c(c_1x) = c(x)$ for all $c_1 \in C$, C a subgroup of X, with c being injective on the set of all cosets $\{Cx\}$. We then call

$$x(k+1) = b[u(k)]\cdot a[x(k)]; \quad y(k) = c[x(k)]$$

a <u>group holomorphic sequential system</u>. We say that such a system is <u>controllable</u> if there exists a positive integer n such that for every x_1, x_2 in X and every integer $m > n$ there exists a control sequence which drives the state from x_1 at $k = 0$ to x_2 at $k = m$. We will say that the system is <u>observable</u> if any two distinct starting states give rise to different output sequences for some choice of input sequence. We call a realization minimal if it is both controllable and observable. We note that the notion of controllability here is stricter than that used earlier. In particular a group holomorphic (but not homomorphic) sequential system can be minimal in the sense of Nerode but not controllable in this sense.

Theorem 2: Let

$$x(k+1) = b[u(k)]\cdot a[x(k)]; \quad y(k) = c[x(k)]$$

be a finite group holomorphic sequential system. Then the set of states reachable from the identity is a subgroup. The set of states indistinguishable from the identity is a subgroup of c and its intersection with the reachable group is normal in the reachable group.

Proof: As we have seen, any holomorphism can be expressed as $\alpha\cdot\beta(\cdot)$ where β is an automorphism on X and α belongs to X. Thus as far as controllability is concerned we can regard holomorphic systems as homomorphic systems with a translation on the input. A glance at the proof of Theorems 5 and 7 of [1] reveals that the reachable set in this case is a subgroup. To prove the second assertion we appeal to Theorem 8 of [2], obvious modifications being made.

Theorem 3: If two minimal* finite group holomorphic sequential systems

$$x(k+1) = b[u(k)]\cdot a[x(k)]; \quad y(k) = c[x(k)]$$

and

$$z(k+1) = g[u(k)]\cdot f[z(k)]; \quad y(k) = h[z(k)]$$

have the same input-output responses from the identity state then their respective state groups X and Z are isomorphic.

Proof: We can write a( ) as $\alpha\cdot\beta(\cdot)$ as noted above. Change notation via $\hat{a} = \alpha^{-1}a(\cdot)$ and $\hat{b}(\cdot) = b(\ )\alpha$. Do the same for f and g. Since $\hat{a}$ and $\hat{f}$ are injective they are permutations and therefore by the finiteness hypothesis for some k and all $\ell=1,2,\ldots$ $\hat{a}^{\ell k}$ is the identity map on X. Likewise $\hat{f}^m$ is the identity for some m. Hence we can find p such that $\hat{a}^{\ell p}(\cdot)$ and $\hat{f}^{\ell p}$ are both identity maps for $\ell = 1,2,\ldots$ By the Nerode theorem there exists a bijection $\rho : X\to Z$ such that the response of system (2) from the state $\rho(x)$ is the same as the response of systems (1) from state x. The systems are assumed to be controllable. Thus for each x_1 we can find a string $u_1,u_2,\ldots u_k$ such that x_1 is the response starting from the identity state after $k = \ell p$ steps.

$$x_1 = \hat{b}(u_1)\cdot\hat{a}(\hat{b}(u_2))\cdot\hat{a}^2(\hat{b}(u_3))\cdot\ldots\cdot\hat{a}^{k-1}(\hat{b}(u_k))\cdot\hat{a}^k(id)$$

thus

* In the sense of Theorem 2.

$$\rho(x_1) = \hat{g}(u_1)\cdot\hat{f}(\hat{g}(u_2))\cdot\hat{f}^2\hat{g}(u_3))\cdot\ldots\hat{f}^{k-1}(\hat{g}(u_k))\cdot\hat{f}^k(id) = z_1$$

Now consider the two machines in states x_2 and $\rho(x_2) = z_2$, respectively. Apply the input string which transfers id to x_1 in k steps. Then we see that since $\hat{a}^k = \hat{f}^h$ = identity map,

$$x_1x_2 = x_1a^k(x_2) = \hat{b}(u_1)\hat{a}(b(u_2))\ldots\hat{a}^k(x_2)$$
$$z_1z_2 = \rho(x_1)\hat{f}^k(\rho(x_2)) = \hat{g}(u_1)\hat{f}(\hat{g}(u_2))\ldots\hat{f}^k(z_2)$$

and so $\rho(x_1x_2) = z_1\cdot z_2 = \rho(x_1)\cdot\rho(x_2)$. Thus ρ is a group homomorphism. Since ρ is one to one and onto it is, in fact, an isomorphism.

We have shown here that even if the assumptions of [1] and [2] are altered substantially (and in a way weakened) it is still possible to establish state space isomorphism theorems. We see that it is not the form of $b(\cdot)$ and $c(\cdot)$ which matter but only the state transition map. This does not, however, render obsolete the results of these papers since we have not given here an analog of the Hankel parameters as one is able to do in the other cases [1], [5]. Neither have we considered questions relating to adjoints [6] and inverses [7], [8] both of which require a fresh approach in this setting.

REFERENCES

1. R.W. Brockett and A.S. Willsky, "Finite Group Homomorphic Sequential Systems," IEEE Trans. on Automatic Control, Vol. AC-17, No. 4, August 1972, pp. 483-490.

2. R.W. Brockett, "System Theory on Group Manifolds and Coset Spaces," SIAM Journal on Control, Vol. No. 2, 1972, pp. 265-284.

3. C.W. Curtis and I. Reiner, Representation Theory of Finite Groups and Associative Algebras, Interscience, N.Y., 1962.

4. M. Hausner and J.T. Schwartz, Lie Groups; Lie Algebras, Gordon Breach, N.Y., 1970.

5. A. Isidori, "Direct Construction of Minimal Bilinear Realizations from Nonlinear Input/Output Maps," IEEE Trans. on Automatic Control, Vol. AC-18, No. 3, December 1973, pp. 626-631.

6. M.A. Arbib and E.G. Manes, "Adjoint Machines, State-Behavior Machines, and Duality," COINS Technical Report 73B-1, Dept. of Computer and Information Science, University of Massachusetts at Amherst, Jan. 1973.

7. A.S. Willsky, Dynamical Systems Defined on Groups: Structural Properties and Estimation, Ph.D. Thesis, Dept. of Aeronautics and Astronautics, M.I.T., Cambridge, Mass., June 1973.

8. A.S. Willsky, "Invertibility Conditions for a Class of Finite State Systems Evolving on Groups," Proc. of the Eleventh Annual Allerton Conference, Univ. of Illinois, Oct. 3-5, 1973.

AUTOMATA IN ADDITIVE CATEGORIES WITH APPLICATIONS TO STOCHASTIC LINEAR AUTOMATA

L. Budach
Mathematics

Humboldt University
East Berlin, Germany

Methods of category theory can be used for different questions of automata theory (see [1] - [5]). The following paper is concerned with the application of categorical algebra to problems which arise in adaptive control of certain technical systems. For this it is of great importance to have available statistical methods for the identification of discrete time, linear, constant, stochastic systems, which will be called stochastic linear automata. To develop these methods it is necessary to mix results of linear algebra, probability-theory, statistics and automata theory. To clear up this intricate situation it is of great use to apply categorical algebra. This will be done in the following abstract. Inspired by the theory of linear automata ([6]) and discrete-time, linear, constant systems ([7]) we develop first a theory of automata over arbitrary additive categories. Especially we obtain a full characterisation of the behaviour of these automata. The observation that the random linear mappings form an additive category is the striking point to apply our main-theorem to stochastic linear automata. It turns out that the identification of stochastic linear automata by input-output experiments becomes possible in substance using statistical methods of autoregressive models.

1. Notations

For basic concepts of categorical algebra we refer the reader to [8]. We consider only additive categories in which finite products always exist.

Let $\underline{C}$ be an additive category. We define two new categories $\underline{C}[[t]]$ and $\underline{C}[t]$. The objects of both are the objects of $\underline{C}$. Moreover we define for two objects X,Y:

$$\underline{C}[[t]](X,Y) = \bigtimes_{i=0}^{\infty} \underline{C}(X,Y)$$

$$\underline{C}[t](X,Y) = \bigoplus_{i=0}^{\infty} \underline{C}(C,Y)$$

So a morphism of X into Y in $\underline{C}[[t]]$ is a sequence $a = (a_o,a_1,\ldots)$ of morphisms in $\underline{C}$, all having the same domain and codomain. If all but a finite number of them are equal to zero, then the morphism belongs to $\underline{C}[t]$. The product and the sum of two morphisms a and b are defined as

$$(ba)_i = b_o a_i + b_1 a_{i-1} + \ldots + b_i a_o$$

$$(b+a)_i = b_i + a_i$$

Obviously the product of two morphisms of $\underline{C}[t]$ belongs to $\underline{C}[t]$. So $\underline{C}[t]$ is a subcategory of $\underline{C}[[t]]$. $\underline{C}$ is a subcategory of $\underline{C}[t]$ if we identify the morphism f of $\underline{C}$ with the sequence $(f,0,0,\ldots)$. If we use for a morphism $a = (a_o,a_1,\ldots)$ the following formal notation

$$a = \sum_{i=0}^{\infty} a_i t^i = a_o + a_1 t + a_2 t^2 + \ldots$$

then morphisms can be multiplied as power series. $\underline{C}[[t]]$ or $\underline{C}[t]$ are called the category of formal power series or category of polynomials over $\underline{C}$ respectively.

A morphism a of $\underline{C}[[t]]$ is called a rational morphism, if $a = a_1 a_2 \ldots a_n$ where the a_i are polynomials over $\underline{C}$ or isomorphisms such that the inverse is a polynomial. All rational morphisms form a subcategory $\underline{C}(t)$ of $\underline{C}[[t]]$. It is easy to see that in the chain

$$\underline{C} \subset \underline{C}[t] \subset \underline{C}(t) \subset \underline{C}[[t]]$$

all categories are additive and all inclusion functors are additive functors.

2. Automata over categories with products.

Let $\underline{C}$ be a category with products and let I be the terminal object of $\underline{C}$. A $\underline{C}$-automaton is a quadruple $S = (X,Y,Z,h)$, consisting of objects X (the input object), Y (the output object), Z (the state object) and a morphism $h\colon Z\times X \to Y\times Z$ called the structure morphism of S. One can define combinatorial automata, delay automata, series-, parallel- and loop-operations of automata in categories with products (see [3]).

A state of a $\underline{C}$-automaton is a morphism $s\colon I \to Z$. We remark that in an additive category any automaton has exactly one state $0\colon 0 \to Z$. A pair (S,s) consisting of an automaton S and a state s of S is to be called an initial automaton.

If $S = (X,Y,Z,h)$ and $S' = (X',Y',Z,h')$ are two automata with the same state object Z, we define the state-product $S\sigma S'$ as follows: the state object is Z, the input object is $X\times X'$, the output object is $Y\times Y'$ and the structure morphism $h\sigma h'$ is equal to $(1_Y\times h')(h\times 1_{X'})$. If (S,s) is an initial automaton then one can define a sequence $S^{(n)}$ of initial automata by $S^{(1)} = S$, $S^{(n+1)} = S^{(n)}\sigma S$. The initial state of these automata is always s.

Let $S = (X,Y,Z,h)$ be an **initial** automaton with initial state s. It defines a morphism

$$F^1(S) = X = I\times X \xrightarrow{s\times 1} X \to Z\times X \xrightarrow{h} Y\times Z \xrightarrow{p_1} Y$$

So we obtain a sequence of morphisms $F^n(S)\colon X^n \to Y^n$ by $F^n(S) = F^1(S^{(n)})$. We set $F(S) = (F^1(S),\ldots)$.

3. Sequential morphisms.

Let $\underline{C}$ be a category with products. We intend to define a new category, the category $\underline{\text{Seq } C}$ of sequential morphisms. The objects of this category are the objects of $\underline{C}$. A sequential morphism $f\colon X \to Y$ is a sequence $f = (f_1^1, f_2^2, \ldots)$ of morphisms $f^n\colon X^n \to Y^n$ in the category $\underline{C}$ satisfying the following condition:

All diagrams

$$\begin{array}{ccc} X^n & \xrightarrow{f^n} & Y^n \\ \downarrow & & \downarrow \\ X^{n-1} & \xrightarrow[f^{n-1}]{} & Y^{n-1} \end{array}$$

(the vertical morphisms are the projections to the first n-1 components) are commutative.

If $f: X \longrightarrow Y$ and $g: Y \longrightarrow Z$ are two sequential morphisms, the product gf is defined by $(gf)^n = g^n f^n$. Notice that there is a canonical isomorphism $\underline{\text{Seq}}\ \underline{C}(X,Y) = \bigtimes_{n=1}^{\infty} \underline{C}(X^n,Y)$, i.e. sequential morphisms can be defined by their corresponding machines.

Proposition. If S is an automaton then F(S) is a sequential morphism.

If S and S' are automata, such that the series-product S'S exists, then F(S'S) = F(S')F(S).

The main observation in the proof of this proposition is that the series-composition and the state-product are related by the following formula: $(S\sigma S')(T\sigma T') = (ST)\sigma(S'T')$. This can be seen by diagram-chasing.

The problem of analysis and synthesis of automata over a given category $\underline{C}$ with products can be formulated as follows:

Give a characterisation of all sequential morphisms $f: X \longrightarrow Y$ such that there exists an automaton S with f = F(S). We shall give a complete solution of this problem for additive categories in the next paragraph.

4. Automata over additive categories.

It is easy to see that sequential morphisms $f: X \longrightarrow Y$ over an additive category $\underline{C}$ are defined by infinite matrices (f_{ij}) with $f_{ij} = 0$ for $j > i$ and $f_{ij}: X \longrightarrow Y$. f is called a constant sequential morphism if for all n, $f_{ij} = f_{i+n,j+n}$.

Proposition and definition. The constant sequential morphisms form a subcategory of $\underline{\text{Seq}}\ \underline{C}$. This category is equivalent to the category $\underline{C}[[t]]$. By this equivalence constant sequential morphisms will be identified with their corresponding power series and this allows us to speak of $\underline{C}[[t]]$ as a subcategory of $\underline{\text{Seq}}\ \underline{C}$. So the notion of a rational or polynomial sequential morphism is defined in the following sense: a sequential morphism is rational or polynomial, if the corresponding power series is rational or polynomial.

Proof. Let $f = (f_{ij}): X \longrightarrow Y$ be a constant sequential function, then f corresponds to $a(t) = \sum_{i=0}^{\infty} f_{i+1,1}t^i$.

Let $S = (X,Y,Z,h)$ be a $\underline{C}$-automaton. The structure morphism $h: Z \oplus X \longrightarrow Y \oplus Z$ is given by the matrix $h = \begin{pmatrix} c & d \\ a & b \end{pmatrix}$.

Proposition. F(S) is a constant, rational sequential morphism.

$$F(S) = d + c(1_Z - at)^{-1}bt$$

Theorem. A sequential function f is an F(S) if and only if f is rational. In particular, all rational morphisms can be expressed in the form $d + c(1_Z - at)^{-1}bt$.

5. Applications to stochastic linear automata.

Definition. A measurable category is a category with products, such that the following conditions are satisfied:

(i) for all objects X,Y $\underline{C}(X,Y)$ has the structure of a measurable space

(ii) the multiplication mapping

$$\underline{C}(X,Y)\times\underline{C}(Y,Z) \longrightarrow \underline{C}(X,Z)$$

is a measurable mapping.

An additive measurable category is an additive category which is measurable such that the mapping

$$-: \underline{C}(X,Y)\times\underline{C}(X,Y) \longrightarrow \underline{C}(X,Y)$$
$$(f,g) \longmapsto f-g$$

is measurable.

Let $\underline{C}$ be a measurable category. If $(S,+,0)$ is a commutative monoid and P is an application of S into the class of all probability-spaces satisfying $\mathcal{P}_0 = \{\emptyset\}$ and $\mathcal{P}_{s+s'} = \mathcal{P}_s \times \mathcal{P}_{s'}$ then we define the category $\underline{C}$ of product-independent random morphisms over $\underline{C}$ as follows: objects are the objects of $\underline{C}$, morphisms $f: X \longrightarrow Y$ are the elements of

$$\underline{C}_{\mathcal{P}}(X,Y) = \bigoplus_{s\in S} \underline{C}_{\mathcal{P}_s}(X,Y)$$

where $\underline{C}_{\mathcal{P}_s}(X,Y)$ is the set of all measurable mappings of $\mathcal{P}_s$ into $\underline{C}(X,Y)$. The product is defined by

$$(gf)(\omega) = \sum g(\omega_1)f(\omega_2)$$

with $\omega \in \mathcal{P}_s$, $\omega_i \in \mathcal{P}_{s_i}$ and $s_1 + s_2 = s$, $(\omega_1,\omega_2) = \omega$. If $\underline{C}$ is an additive measurable category, then $\underline{C}_{\mathcal{P}}$ is additive. $\underline{C}_{\mathcal{P}}$-automata can be interpreted as follows: they are stochastic $\underline{C}$-automata, which satisfy the condition, that the transition and output-function at time $t + 1$ are as stochastic events independent of these functions at time t.

Let us consider an example. Let $\underline{C}$ be the category of finite dimensional vector-spaces over the field of real numbers. The σ-algebra of Borel-sets makes $\underline{C}(S,Y)$ to a measurable space. The multiplication and subtraction are measurable mappings. Thus $\underline{C}_{\mathcal{P}}$ is an additive category, the category of random linear mappings. What we mean exactly if we speak of a stochastic linear automaton is an automaton over the category $\underline{C}_{\mathcal{P}}$.

Our theorem gives us a full characterization of the behavior of stochastic linear automata. Using statistical methods of autoregressive models it is now possible to identify linear stochastic automata by input-output experiments. Details will be given in a forthcoming paper.

REFERENCES

[1] L. Budach, H.-J. Hoehnke: Über eine einheitliche Begründung der Automatentheorie, mimeographed seminar notes, 1969/70.

[2] L. Budach: Kohomologie von abstrakten Automaten, mimeographed seminar notes 1969/70.

[3] L. Budach, H.-J. Hoehnke: Automaten und Funktoren, Akademieverlag Berlin (in press).

[4] L. Budach: Sequentielle Gruppen, Mitt. der Math. Gesell. der DDR, 23, 12-23 (1973).

[5] L. Budach: Group objects in the category of automata, Proceedings of the conference "Algebraic Methods in Automata-Theory" Szeged, 1973 (in press).

[6] A. Gill: Linear sequential circuits, McGraw-Hill Inc., New York, 1966.

[7] R. E. Kalman: Lectures on Controllability and Observability, Centro Internaxionale Mathematico Estivo, Bologna, 1968.

[8] S. Mac Lane: Categories for the Working Mathematician, Springer-Verlag, 1972.

[9] A. Heller: Probabilistic Automata and Stochastic Transformations, Math. Sys. Th., I, No. 5, 1967.

THE ALGEBRAIC THEORY OF RECURSIVE PROGRAM SCHEMES

R. M. Burstall
Department of Machine Intelligence
Edinburgh University
Edinburgh, Scotland

J. W. Thatcher
Mathematical Sciences Department
IBM T. J. Watson Research Center
Yorktown Heights, New York

This abstract concerns the authors' research in progress. It does not abstract a completed work. We have a principal example, recursive monadic program schemes, which is described intuitively and formally below. We believe that the proper formulation of this example will have general applications and we are still seeking that formulation. It lies somewhere in the neighborhood of a (bi-)category whose hom-sets are cocomplete categories and whose composition is functorial and preserves filtered colimits. Call such a category a $\underset{\sim}{C}$-category. The hope is that the construction of the category of recursive monadic program schemes generalizes, beginning with an arbitrary cocomplete category, yielding a $\underset{\sim}{C}$-category

Some of the ideas for this work grew out of the collaboration of one of us (J.T.) with J. Goguen, E. Wagner and J. Wright [3]. We particularly acknowledge subsequent efforts by Goguen, Wagner and Wright to help us overcome obstacles we found in attempting to carry forward our general program. Goguen's outline of the formulation of RMPS resulting from joint discussions helped greatly in writing this abstract.

This work was stimulated by a seminar given by W. P. de Roever at the Department of Machine Intelligence, Edinburgh University, while J. T. was visiting there on sabbatical from IBM with U. K. Science Research Council sponsorship. We are both grateful to Edinburgh University, IBM, and SRC for their support.

THE PRINCIPAL EXAMPLE, INFORMALLY. A monadic recursive programming situation consists of a set of programs each having a name and in which calls to any of the programs in the set can occur. Being monadic, the operations of the programs (and the programs themselves) are analyzed only insofar as they effect the entire memory state.

There are many ways of formulating the semantics of such recursive programs. The distinction between computational (operational) semantics and denotational (mathematical) semantics is drawn. The interpretive approach of Landin's SECD machine [4] and the computational models of de Bakker and de Roever [1,2] seem to fall in the former classification, the fixed point equations of Scott and Strachey [9] are in the latter and somewhere, related to both, is the "copy rule" of, say ALGOL 68 report [10].

That is the principal example. The mathematical formulation for recursive programs and their semantics that we seek will at least blur, if not eliminate, the distinction between computational and denotational semantics. Nivat [7] takes a similar approach both in its algebraic character (though not categorical) and uniformity.

NOTATIONAL PRELIMINARIES. (1) We use ordinal notation for the non-negative integers; $n = \{0,1,\cdots,n-1\}$, $\omega = \{0,1,2,\cdots\}$. (2) For a category $\underline{\underline{C}}$, $|\underline{\underline{C}}|$ denotes the objects of $\underline{\underline{C}}$. (3) $\underline{\underline{\underline{Set}}}$ is the category of sets and functions. (4) A directed graph consists of sets V (vertices), E (edges), and functions $\partial_i\colon E \to V$ $(i=0,1)$. $G(v_0,v_1) = \{e \mid e\partial_i = v_i\}$. $\underline{\underline{G}}^*$ is the path category of G. (5) $\underline{\underline{G}}_{**}$ is the category of reduced protected bi-pointed graphs, i.e., directed graphs G, with distinguished vertices $\underline{in}$ and $\underline{out}$ (preserved by the morphisms) such that:

(a) (reduced) $\underline{\underline{G}}^*(\underline{in},v) = \emptyset$ iff $\underline{\underline{G}}^*(v,\underline{out}) = \emptyset$ for all vertices $v \notin \{\underline{in},\underline{out}\}$.

(b) (protected) $G(\underline{out},v) = G(v,\underline{in}) = \emptyset$ for all vertices v.

THE CATEGORY OF MONADIC PROGRAM SCHEMES. Let Ω be a fixed set (of operation symbols). The category of monadic program schemes, $\underline{\underline{MPS}}_\Omega$ has vertical and horizontal structure. The vertical structure is a skeletal subcategory of the comma category $(U\downarrow \ulcorner\Omega\urcorner)$ where $U\colon \underline{\underline{G}}_{**} \longrightarrow \underline{\underline{Set}}$ forgets everything except the edges and $\ulcorner\Omega\urcorner\colon \underline{1} \longrightarrow \underline{\underline{Set}}$ just picks out Ω from $|\underline{\underline{Set}}|$. (See Mac Lane [6]) So an object of $\underline{\underline{MPS}}_\Omega$ is a bipointed graph with edges labeled from Ω and a morphism is a bi-pointed graph morphism which "preserves labels."

The horizontal structure of $\underline{\underline{MPS}}_\Omega$ is a monoid (category with one object). $P_0 ; P_1$ is the reduced labeled bi-pointed graph obtained by identifying the input node of P_1 with the output note of P_0.

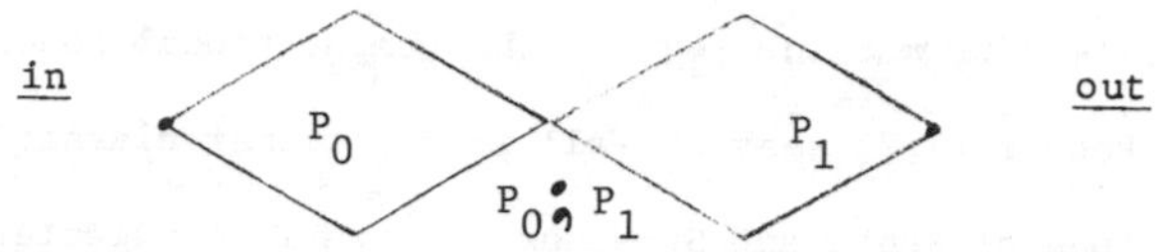

Define $\perp$ to be the "totally undefined" scheme with two distinct nodes in and out and no edges. Then by reduction, $P ; \perp \;=\; \perp ; P = \perp$. With 1 taken to be the degenerate program scheme with one node and no edges (the one node distinguished as both in and out) $1 ; P = P ; 1 = P$.

FACTS. (1) $\underline{\underline{MPS}}_\Omega$ is countably cocomplete.

(2) $\perp$ is initial.

(3) $; \colon \underline{\underline{MPS}}^2_\Omega \longrightarrow \underline{\underline{MPS}}_\Omega$ is a functor cocontinuous with respect to filtered colimits.

Define $P_1 \subseteq P_2$ iff there exists a vertical morphism from P_1 to P_2 in $\underline{\underline{MPS}}_\Omega$. Fact (1) looks like it says that this order has joins but it doesn't just because $\subseteq$ is not a partial order. For example $P_1 \subseteq P_2$

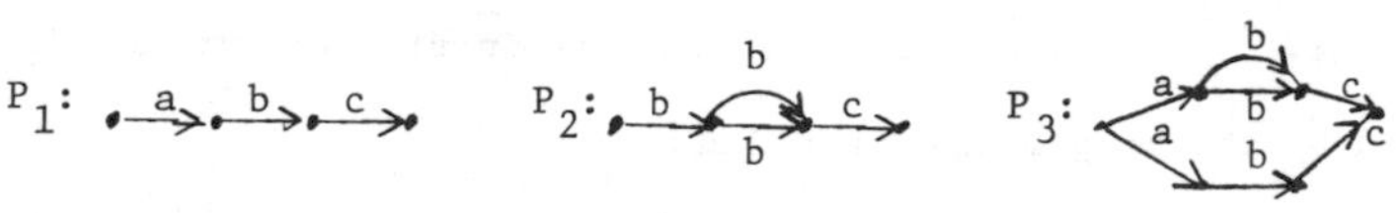

and $P_2 \subseteq P_1$ by the obvious scheme morphisms, say, $f_1\colon P_1 \longrightarrow P_2$ and $f_2\colon P_2 \longrightarrow P_1$. But in $\underline{\underline{MPS}}_\Omega$ all the diagrams,

$$\underset{\bullet}{P_2} \xrightarrow{f_2} \underset{\bullet}{P_1} \qquad\qquad \underset{\bullet}{P_1} \xrightarrow{f_1} \underset{\bullet}{P_2} \qquad\qquad \underset{\bullet}{P_1} \quad \underset{\bullet}{P_2}$$

have colimits which are P_1, P_2 and P_3 respectively. With only an order relation between schemes we could not capture this situation; here the vertical category structure is essential.

But if we cut $\mathbf{MPS}_\Omega$ down to the full subcategory of determinate program schemes (no two edges with the same source have the same label), then all morphisms are injections; in fact, $\subseteq$ is a partial order. This structure is essentially Scott's lattice of flow diagrams [8] without $\top$, i.e., joins are not defined for all sets and can be considered to be defined for directed sets only.

THE ALGEBRAIC THEORY OF RECURSIVE PROGRAM SCHEMES. Let Σ be a fixed set (of operator symbols) disjoint from ω. Also let $\mathbf{MPS}'_\Omega$ be the subcategory of (the vertical structure of) $\mathbf{MPS}_\Omega$ excluding the degenerate scheme, 1. The theory of recursive program schemes, $\mathbf{RMPS}_\Sigma$ has objects ω and vertical structure, $\mathbf{RMPS}_\Sigma(m,n) = (\mathbf{MPS}'_{\Sigma\cup n})^m$. Thus objects are m-tuples of non-degenerate $(\Sigma\cup n)$-schemes (view n as a set of program names, $z_0, z_1, \ldots, z_{n-1}$) and morphisms are m-tuples of scheme morphisms.

Horizontal structure is substitution (the copy rule). Given $R\colon m \longrightarrow \mathbf{MPS}'_{\Sigma\cup n}$ and $R'\colon n \longrightarrow \mathbf{MPS}'_{\Sigma\cup p}$, $R\circ R'$ is the m-tuple of (reduced) schemes obtained from simultaneously substituting scheme iR' for each occurrence of an edge labeled i in all the components of R $(i \in n)$.

$\bot_{m,n} \in \mathbf{RMPS}_\Sigma(m,n)$ is the m-tuple of $\bot$'s. Although this $\bot$ (we can drop the subscripts) inherits its initiality from $\mathbf{MPS}_{\Sigma\cup n}$ things are now more interesting because $P\circ\bot$ is not necessarily $\bot$.

FACTS. (1) Each $\mathbf{RMPS}_\Sigma(m,n)$ is countably cocomplete and the colimits respect (preserve) tupling.

(2) $\bot$ is initial in $\mathbf{RMPS}_\Sigma(m,n)$.

(3) $\circ\colon \mathbf{RMPS}_\Sigma(m,n) \times \mathbf{RMPS}_\Sigma(n,p) \longrightarrow \mathbf{RMPS}_\Sigma(m,p)$ is a functor cocontinuous with respect to filtered colimits.

(4) $\mathbf{RMPS}_\Sigma$ is an algebraic theory (See Lawvere [5]).

Notice that to get lattice structure through determinate recursive schemes, we would have to require that any edge labeled with a program variable be the only edge with its source, else nondeterminate schemes could result from substitution.

SEMANTIC CONSIDERATIONS. The principal example is a $\underset{\sim}{C}$-category. In such categories and in $\underline{\underline{RMPS}}_\Sigma$ in particular we can take fixed points:

FACT. For $R \in \underline{\underline{RMPS}}_\Sigma(n,n)$ there exists $\mu_R \in \underline{\underline{RMPS}}_\Sigma(n,n)$ such that $R \circ \mu_R = \mu_R$ and if $R \circ S = S$ then there exists a vertical $f: \mu_R \longrightarrow S$.

For such $\underset{\sim}{C}$-categories and $\underline{\underline{RMPS}}_\Sigma$ in particular, there is internal semantics. Internal semantics is fixed points which is boring in $\underline{\underline{MPS}}_\Omega$ because $\mu_P = \perp$ but very interesting in $\underline{\underline{RMPS}}_\Sigma$ because it is essentially the syntactic "copy rule" yielding for each $R: n \rightarrow n$ an n-tuple of infinite flow diagrams with no program variables. Other kinds of semantics are found as $\underset{\sim}{C}$-functors from $\underline{\underline{RMPS}}_\Sigma$ to other $\underset{\sim}{C}$-categories.

For the principal example, consider the algebraic theory $\underline{\underline{Sub}}_\Sigma$ where $\underline{\underline{Sub}}_\Sigma(m,n) = (\wp(\Sigma \cup n)^*)^m$ -- m-tuples of sets of strings over Σ together with n variables. Vertical structure is the lattice structure on $(\wp(\Sigma \cup n)^*)^m$ and horizontal structure is string substitution. (When $S \in \underline{\underline{Sub}}_\Sigma(n,n)$ is finite, it is just a context-free grammar with n nonterminals. $i\mu_S$ is the language generated by S with start symbol i.) The natural map Comp: $\underline{\underline{RMPS}}_\Sigma \longrightarrow \underline{\underline{Sub}}_\Sigma$ is indeed a ($\underset{\sim}{C}$-) functor, where (R)Comp is the n-tuple of sets of strings of labels from paths from in to out in the components of R. This is but one of many possible interpretations of (semantics for) recursive program schemes and by ($\underset{\sim}{C}$-) functorality, it agrees with the internal semantics. Another (and more familiar) semantics has as target the ($\underset{\sim}{C}$-) category of relations yielding the conventional monadic fixed point semantics.

REFERENCES

[1] de Bakker, J. W., "Recursive procedures," Math. Cent. Tracts 24, Amsterdam, 1971.

[2] de Roever, W.P., "Operational and mathematical semantics for recursive polyadic program schemes," Proceedings, Symposium on Mathematical Foundations of Computer Science, Computing Research Centre, Slovak Academy of Sciences, Bratislava, Czechoslovakia, 1973.

[3] Goguen, J.,Thatcher, J., Wagner, E., and Wright, J., "A junction between computer science and category theory," I,II Basic definitions and examples; III,IV Algebraic theories and structures; V Initial factorizations and dummy variables. IBM Research Reports: I Part 1, RC-4526, September 1973; others to appear.

[4] Landin, P. J., "The mechanical evaluation of expressions," Comput. J. 6 (1964) 308-320.

[5] Lawvere, F. W., "Functorial semantics for algebraic theories," Proceedings, Nat'l Acad. Sci. 50 (1963) 869-872.

[6] Mac Lane, S., Category Theory for the Working Mathematician, Springer-Verlag, 1971.

[7] Nivat, M., "Languages algebriques sur le magma libre et semantigue des schemas de programme," in Automata, Languages and Programming (M. Nivat, ed.) North Holland (1972) 293-308).

[8] Scott, D., "The lattice of flow diagrams," Oxford University Computing Laboratory Technical Monograph PRG-3, 1970.

[9] Scott, D. and Strachey, C., "Towards a mathematical semantics for computer languages," Proceedings, Symposium on Computers and Automata, Polytechnic Institute of Brooklyn 21, 1971.

[10] van Wijngaardsn, A. (ed.) "Report on the algorithmic language ALGOL 68." Math. Cent. Report MR 101, Amsterdam, 1969.

REALIZATION IS CONTINUOUSLY UNIVERSAL

Lee A. Carlson
Information Sciences

University of Chicago
Chicago, Ill. 60637, U.S.A.

Valparaiso University (on leave)
Valparaiso, Ind. 46383, U.S.A.

The adjunction between minimal realization and behavior has been displayed for discrete-time machines by Goguen [3, 4], and the minimal realization of continuous-time invariant linear differential systems has been developed by Kalman and Hautus [5] and Bainbridge [1] by reduction to a discrete-time problem. This paper, following the framework of Goguen [3], establishes the minimal realization adjunction and associated results for continuous-time invariant linear differential systems working directly in continuous time. System objects are described in terms of sheaves of (topological) vector spaces - sheaves embody a natural structure for specifying a self-consistent collection of system observations - and input-output maps are described by sheaf maps induced by linear transformations - the traditional matrices are not used in order to avoid the complications associated with basis choice. It is hoped that the concrete construction of this paper will subsequently generalize to other differential systems.

The methods of the paper are naive, constructing objects and maps pointwise. The reader is refered to MacLane [6] for categorical terminology and Seebach et. al. [7] for background in sheaves. Familiarity with classical elementary linear system theory [2] is assumed for motivation and explication.

System objects are obtained by the following sheaf construction. Let $\underline{\underline{T}}$ denote continuous time, the reals with topology $\{O^t = \langle -\infty , t\rangle\}$. For a real vector space V, the corresponding sheaf $\underline{V}$ contains, for each O^t, the vector space $\underline{V}(O^t)$ of piecewise C^∞ functions* v^t, continuous from the left with left bounded support, and the restriction maps given pointwise by $v^t|_{t'} = v^{t'}$ in $\underline{V}(O^{t'})$. Given a linear transformation $a:V \to W$, write $\underline{a}:\underline{V} \to \underline{W}$ for the pointwise induced sheaf map. $S_t:\underline{V} \to \underline{V}$ will denote the time shift (by t) operator acting pointwise on $\underline{V}$, $D:\underline{V} \to \underline{V}$ the differential operator (the derivative where it exists, extended by left continuity), and $e:\underline{V} \to V$ denotes limit evaluation ($e(v^{t'}) = \lim_{t \to t'} v^{t'}(t)$). $\underline{V}(O^t)$ admits concatenation of functions on $\langle t, t'\rangle$ giving functions in $\underline{V}(O^{t'})$ (again using left continuity).

A continuous-time invariant linear differential machine $M = \langle X, S, Y, \delta, \lambda\rangle$ is given by vector spaces X, S, Y and linear transformations $\delta:X + S \to S$, $\lambda:S \to Y$. Morphisms $\langle a, b, c\rangle:M \to M'$ of machines are tripples of linear transformations $a:X \to X'$, $b:S \to S'$, $c:Y \to Y'$ which preserve the actions of δ, λ in M'. Componentwise composition gives the category $\underline{\underline{M}}\text{ach}$ of machines (differential systems). For each machine M the solution of the differential equation $Ds = \delta(x,s)$

* defined on O^t

guarantees the existence of an equalizer ([6]) $\mathfrak{s}$

$$\underline{X} \overset{\mathfrak{s}}{\rightarrow} \underline{X}+\underline{S} \underset{DP}{\overset{\delta}{\rightrightarrows}} \underline{S}$$

(P is the projection to $\underline{S}$) in the category $\underline{\underline{S}}$heaves, which is used to obtain the solution map $\delta^+(x^t) = s^t$ which gives the state trajectory corresponding to each input function. M is reachable iff $e\delta^+$ is surjective. Recall that it is equivalent for $e\delta^+$ restricted to smooth x^t to be surjective [2]. $\underline{\underline{M}}$ denotes the subcategory of reachable machines and morphisms $<a, b, c>$ with surjective a.

A continuous-time invariant behavior (not necessarily differential) is a sheaf map $f:\underline{X} \to \underline{Y}$ linear on each $\underline{X}(0^t)$ such that $S_t f = f S_t$ for all t. Morphisms $<a, c>:f \to f'$ are such that $f'\underline{a} = \underline{c}f$. This gives the category $\underline{\underline{B}}$eh of behaviors, and the external behavior functor $E:\underline{\underline{M}}ach \to \underline{\underline{B}}eh$ is obtained by setting $E(M) = \underline{\lambda}\delta^+$, $E(<a,b,c>) = <a,c>$.

For a behavior f in $\underline{\underline{B}}$eh, the Nerode equivalence relation is given by $x^t \sim x^{t'}$ iff f agrees, except for time shift, on each common continuation (using the concatenation operation and requiring agreement on the half-closed interval $[t, t+h>$). The equivalence gives a quotient map $q:\underline{X} \to S_f$ to a "natural" linear state space for any linear behavior. However, the construction is more general than necessary, e.g., S_f exists for f the unit delay operator. In order to restrict attention to differential behaviors an "impulse response" map is used: write 'q' for the function mapping $\underline{X}(0^{t'}) \to \underline{S}_f(0^{t'})$ given by $'q'(x^{t'})(t) = q(x^{t'}|_t)$. The impulse responce is to be defined mapping $X \to S_f$ by $I(x) = \lim_{h \to 0} eD('q'(x \cdot 1_o^h))$ where 1_o^h is the unit step function at $0<h$ defined on $<-\infty , h>$.

Linear differential behaviors are those for which:

A. $D'q'(x^t) = 'q'D(x^t)$ for smooth x^t.

B. Each Nerode equivalence class contains a smooth x^t.

C. $I:X \to S_f$ is defineable as above.

D. $qD(x^t) = I(ex^t)$ whenever $q(x^t) = 0$, x^t smooth.

The specification of smooth x^t simplifies the notation and formulation of the properties. Property C is explicated by determining that $I(x) = q(Dx^t)$ where $x^t \sim 0$ is smooth and $ex^t = x$. All four properties are then interpreted in terms of the classical $s(t) = \int_{-\infty}^{t} \exp(A(t-\tau))Bx(\tau)$.

The subcategory of $\underline{\underline{Beh}}$ containing linear differential behaviors and morphisms <a,c> with surjective a is denoted $\underline{\underline{B}}$. These need not be behaviors of finite dimensional machines, though each may be realized: The realization functor $N:\underline{\underline{B}} \to \underline{\underline{M}}$ is constructed by defining $\delta_f(x, qx^t) = qD(x^t) - I(ex^t) + I(x)$, where x^t is smooth, and $\lambda_f q(x^t) = ef(x^t)$, obtaining $N(f) = \langle X, S_f, Y, \delta_f, \lambda_f\rangle$ in $\underline{\underline{M}}$. The corresponding differential equation may be solved to obtain $\delta_f^+(x^t) = 'q'(x^t)$. For $\langle a,c\rangle:f \to f'$ define $b^*:S_f \to S_{f'}$ by $b^*q(x^t) = q'(\underline{a}x^t)$ to obtain $N(\langle a,c\rangle) = \langle a,b^*,c\rangle:M_f \to M_{f'}$. It is then easily verified that $EN(f) = f$, so that N is a realization functor.

A "state reduction" natural transformation ρ can be defined by taking $\rho_M:M \to NEM$ to be the identities on X and Y, and on the states $q_M:S \to S_f$ given by $q_M(s) = q(x^t)$, where $e\delta^+(x^t) = s$. This verifies that the Nerode realization of a finite dimensional machine is finite dimensional, and - more importantly - allows the extension of Goguen's results [3] (by essentially identical arguments) to continuous time machines: behavior is left adjoint to realization, reduced machines are a reflexive subcategory of reachable machines, and realizations are minimal iff reachable and reduced.

REFERENCES

1. Bainbridge, E. S., A Unified Minimal Realization Theory, with Duality, Dissertation, Department of Computer and Communication Sciences, U. of Mich. (1972)
2. Chen, C. T., Introduction to Linear System Theory, Holt, Rinehart and Winston, (1970)
3. Goguen, J. A., "Realization is Universal," Math. Sys. Th. 6 (1973) 359-374
4. ......, "Minimal Realization of Machines in Closed Categories," Bull. Amer. Math. Soc. 78 (1973) 777-783
5. Kalman, R. E., and Hautus, M. L. J., "Realization of Continuous-time Linear Dynamical Systems: Rigorous Theory in the Style of Schwartz," NRL-MRC Conference: ODE, Acad. Press (1972) 151-164
6. MacLane, S., Categories for the Working Mathematician, Springer-Verlag (1971)
7. Seebach, J. A., Jr., Seebach, L. A., and Steen, L. A., "What is a Sheaf," Amer. Math. Monthly, MAA 77 (1970) 681-703

DIAGRAM-CHARACTERIZATION OF RECURSION

Hartmut Ehrig (FB 20)
Wolfgang Kühnel (FB 3)
Michael Pfender (FB 3)

Technical University of Berlin, D-1 Berlin 12, GERMANY

The aim of the present paper is a diagrammatic hence element-free description of μ-recursion and while-loops. This constitutes the base for an algebraic treatment of recursively defined functions on the syntactical level of suitably enriched algebraic theories in Lawvere's sense. For related approaches see [1], [2], [4], [8], [9]. In § 1 μ-recursion and while-loops are characterized by commutative diagrams in the category of sets. All concepts are reduced to the well-order of the natural numbers. § 2 gives a topos-theoretical generalization and § 3 an equational characterization of μf and an idea of further development.

§ 1 CHARACTERIZATION IN SETS

1.1 Definition: Given a function $f: A \times N \longrightarrow 2$ (N = natural numbers, $2 = \{0, 1\}$), $\mu f: A' \longrightarrow N$ is given by:

(i) $A' = \{a \in A / \exists\, n \in N: f(a, n) = 1\} \subseteq A$,

(ii) $\mu f(a) = \min\{n \in N / f(a, n) = 1\}$ for $a \in A'$.

The choice of 1 instead of 0 in this definition is motivated by the generalization in § 2, where the truth value "true" is the basic, "false" only a derived one.

<u>1.2 Proposition:</u> $\mu f: A' \to N$ is characterized by (1), (2), (3) :

(1)

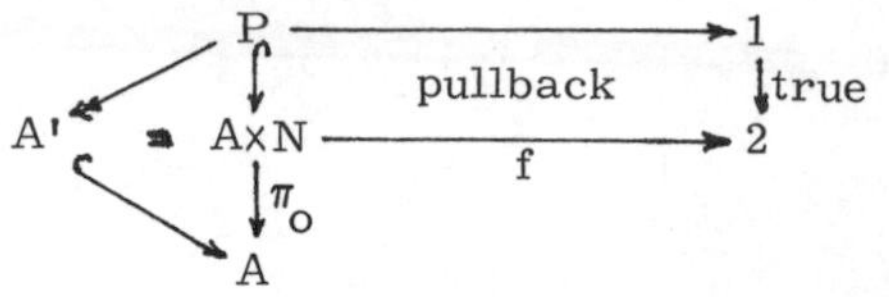

i.e. $P = f^{-1}(1)$, $A' = \pi_o(P) \subseteq A$, true being defined by true(0) = 1,

(2)

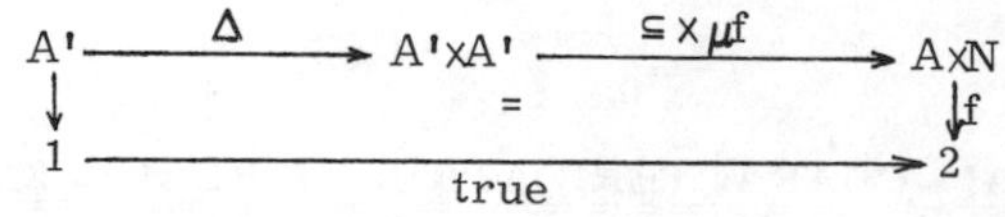

(3)

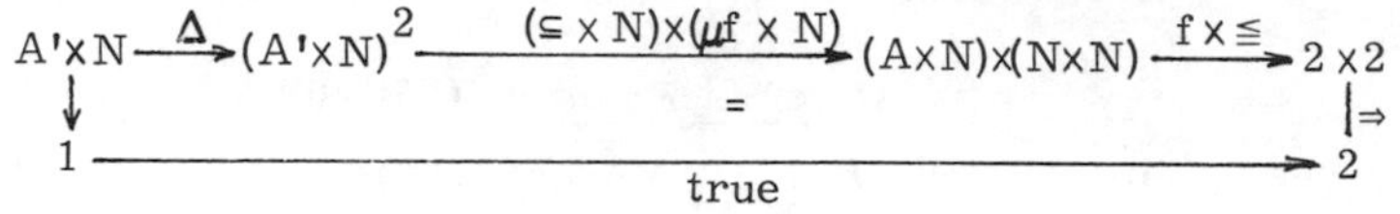

<u>1.3 Corollary:</u> Taking $f := ev: 2^N \times N \to 2$ in 1.2, we get a characterization of the μ-operator (well-ordering) $\mu = \mu(ev): P^+N \to N$ of the natural numbers, defined by $\mu(\chi) = \min\{n \in N / \chi(n) = 1\}$.

<u>1.4 Proposition:</u>

(i) For $f: A \times N \to 2$ μf is given by

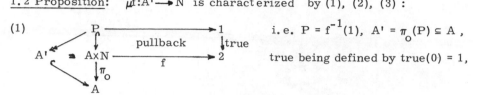

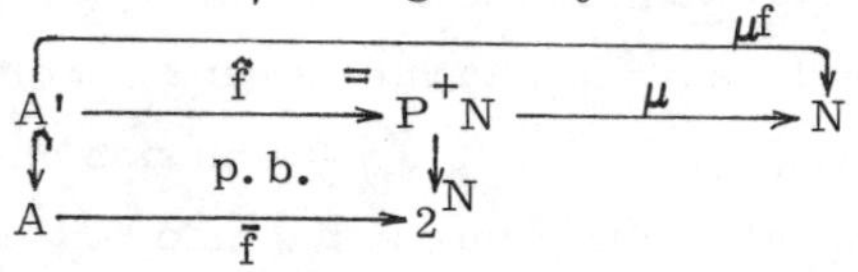

(1)

with $A \xrightarrow{\bar{f}} 2^N$ adjoint to $A \times N \xrightarrow{f} 2$.

(ii) $\mu: P^+N \to N$ is a retraction of the singleton-map $\{\ \}^+: N \to P^+N$.

(iii) For each $g: A \to N$ there is at least one $f: A \times N \to 2$ satisfying $\mu f = g$.

We now describe the while-concept being a well-known recursion-concept in programming languages. It will appear in the form "until p do f", because we prefer 1 instead of 0 (see 1.1).

<u>1.5 Definition:</u> Given functions $f: A \to A$, $p: A \to 2$ (predicate), $\text{until}(p, f): A' \to A$ is defined by:

(i) $A' = \{a \in A / \exists\ n \in N: pf^n(a) = 1\} \subseteq A$,

(ii) $\text{until}(p, f)(a) = f^m(a)$ with $m = \min\{n \in N / pf^n(a) = 1\}$ for all $a \in A'$.

<u>1.6 Proposition</u>: $\mathrm{until}(p,f): A' \longrightarrow A$ is characterized by (1), (2), (3), (4) :

(1)

$$\begin{array}{ccccccc} & & P & \longrightarrow & & & 1 \\ & \swarrow & \downarrow & & \text{p.b.} & & \downarrow \text{true} \\ A' & \subset = & A\times N & \xrightarrow{f^{\bullet}} & A & \xrightarrow{p} & 2 \\ & \searrow & \downarrow \pi_o & & & & \\ & & A & & & & \end{array}$$

with $f^{\bullet}(a,n) := f^n(a)$

(2) A_1 defined by

$$\begin{array}{ccccc} A_1 & \longrightarrow & & & 1 \\ \downarrow & & \text{p.b.} & & \downarrow \text{true} \\ A' & \hookrightarrow & A & \xrightarrow{p} & 2 \end{array}$$

i.e. $A_1 := p^{-1}(1)$

satisfies

$$\begin{array}{ccc} A_1 & \hookrightarrow & A' \\ \downarrow & = & \downarrow \mathrm{until}(p,f) \\ A' & \hookrightarrow & A \end{array}$$

i.e. $\mathrm{until}(p,f)|_{A_1} = \text{incl.}$

(3) f restricts to $A_o \xrightarrow{f_o} A'$ with $A_o := p^{-1}(0) \cap A'$ and

$$\begin{array}{ccc} A_o & \xrightarrow{f_o} & A' \\ \downarrow & = & \downarrow \mathrm{until}(p,f) \\ A' & \xrightarrow[\mathrm{until}(p,f)]{} & A \end{array}$$

i.e. $\mathrm{until}(p,f)|_{A_o} = \mathrm{until}(p,f) \circ f_o$

(4)

$$\begin{array}{ccc} A' & \xrightarrow{\mathrm{until}(p,f)} & A \\ \downarrow & = & \downarrow p \\ 1 & \xrightarrow[\text{true}]{} & 2 \end{array}$$

<u>Remark</u>: In fact, (4) follows from (1) - (3), and (2) - (3) correspond to a similar condition of [1] for complete lattices.

<u>1.7 Proposition</u>: The three concepts of μ-recursion, while-loops and well-order of the natural numbers can be expressed by each other via the following equations:

(i) $\mu f = \mu \circ \hat{f}$ for all $f: A\times N \longrightarrow 2$ (cf. 1.4 (i))

(ii) $\mathrm{until}(p,f)(a) = f^{\mu(pf^{\bullet})(a)}(a)$ for each $a \in A'$, for all $f: A \longrightarrow A$ and $p: A \longrightarrow 2$, in diagram form:

$$\begin{array}{ccc} A' & \xrightarrow{\mathrm{until}(p,f)} & A \\ \Delta \downarrow & = & \uparrow f^{\bullet} \\ A' \times A' & \xrightarrow[\subseteq \times [\mu(pf^{\bullet})]]{} & A\times N \end{array}$$

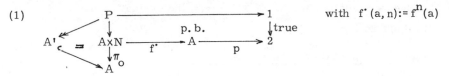

(iii) $\mu(\chi) = \pi_1 \circ \mathrm{until}(ev, 2^N \times s)(\chi, 0)$ for each $\chi \in P^+N$, $s:N \to N$ being the successor function and $ev:2^N \times N \longrightarrow 2$ the evaluation map, in diagram form:

$$\begin{array}{ccc} P^+N & \xrightarrow{\quad \mu = \mu(2^N \times N \xrightarrow{ev} N) \quad} & N \\ \downarrow{\scriptstyle (\subseteq, 0)} & = & \uparrow{\scriptstyle \pi_1} \\ (2^N \times N)' & \xrightarrow[\quad \mathrm{until}(ev, 2^N \times s) \quad]{} & 2^N \times N \end{array}$$

with $(2^N \times N)' = \{(\chi, n) \in 2^N \times N \ / \ \exists\ m \in N: \ \chi(s^m(n)) = 1\}$ definable as the pullback (1) in 1.6 .

§ 2 μ-RECURSION IN TOPOI

All of the preceding concepts are definable in topoi with natural number object N. Intended are results on existence and uniqueness of μ, μf, until(p, f) in topoi and on equivalence of these different concepts. In this paper we use reflexivity and antisymmetry of the order $\leq : N \times N \longrightarrow 2$ (cf. [10]), moreover $0 = \min N$ and compatibility of $\leq$ with the successor function $s:N \longrightarrow N$.

<u>2.1 Definition</u>: For $f:A \times N \longrightarrow 2$ a morphism $\mu f:A' \longrightarrow N$ satisfying (1) - (3) of 1.2 is called μ-recursion of f. A morphism $\mu:P^+N \longrightarrow N$ satisfying (1) - (3) for $f = ev:2^N \times N \to 2$ (cf. 1.3) is called a μ-operator of the NNO N. The definition of P^+N by (1) is well-known in topos-theory.

<u>2.2 Proposition</u>: Given $f:A \times N \longrightarrow 2$, there is at most one morphism $\mu f:A' \longrightarrow N$ satisfying (1) - (3) of 1.2 . Hence the μ-recursion of f and especially the μ-operator of the NNO in 2.1 are uniquely determined.

<u>2.3 Theorem</u>: For $f:A \times N \longrightarrow 2$ the μ-recursion μf is given by $\mu f = \mu \circ \hat{f}$ with $\hat{f}$ defined by the pullback in 1.4 (i) .

<u>Remark</u>: The reduction of arbitrary μ-recursion to the μ-operator of the NNO corresponds to the reduction of primitive recursion to simple recursion (defining universal property of the NNO) in [3] . Both reductions and proofs lean heavily on the cartesian closure structure of topoi.

<u>2.4 Corollary</u>:

(i) The μ-operator of N is a retraction of $\{\ \}^+:N \longrightarrow P^+N$, defined as the codomain-restriction of the singleton-map $\{\ \}: N \longrightarrow 2^N$.

(ii) For each $g:A \longrightarrow N$ there is at least one $f:A \times N \longrightarrow 2$ satisfying $\mu f = g$. f is defined as the adjoint morphism of $\bar{f} := A \xrightarrow{g} N \xrightarrow{\{\ \}} 2^N$.

<u>2.5 Theorem:</u> Given $f:A \to A$ and $p:A \to 2$, until(p, f) can be defined by diagram (ii) in 1.7 provided that $\mu(pf^{\bullet})$ satisfies the axioms (1) - (3) in 1.2 where $f^{\bullet}$ is the adjoint of $\tilde{f}:N \to A^{A}$ which in turn is the unique morphism satisfying

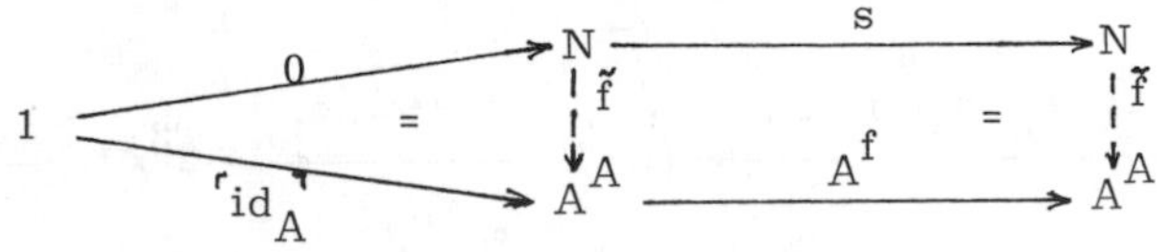

(universal property "simple recursion" of the NNO in a topos (cf. [3])).
Remark: For the construction of $f^{\bullet}$ Joyal's condition of N to be a free monoid relative actions is sufficient (cf. [5]) :

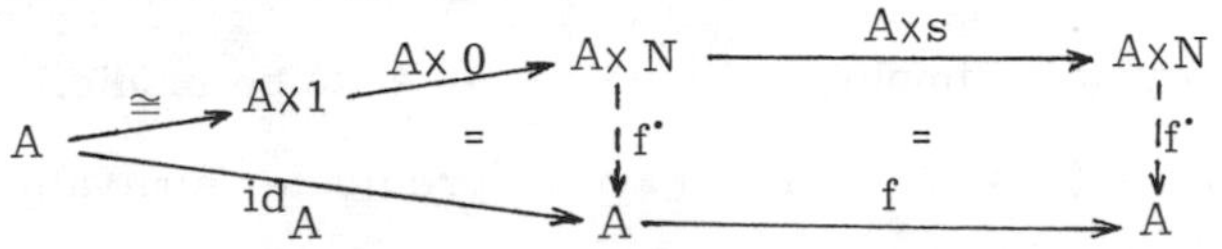

<u>2.6 Remark:</u> By 2.5 we can construct until(p, f) using only the existence of $\mu(pf^{\bullet})$, and by 2.3 this can be defined by μ. These relations correspond to (i) and (ii) of 1.7 . There is the natural question whether this can be proved for (iii) of 1.7, too. This would imply the equivalence of all three concepts. We suppose that the answer is 'yes' at least for Boolean topoi.

§ 3 EQUATIONAL CHARACTERIZATION OF RECURSION, FURTHER DEVELOPMENT

As mentioned in the introduction we want to interpret recursive program schemes as morphisms within suitable 'first-order logical theories', freely generated over the basic function and predicate symbols of the class of program schemes under consideration. These theories will be heterogeneous algebraic theories in the sense of Lawvere ([6]) which are in the same time similar to the logical categories in the sense of [7].
To incorporate recursion, for each $f:A \times N \to 2$ a $\mu f:A' \to N$ should then be freely adjoint, and (1) - (3) of 1.2 be guaranteed by a suitable factorization of the theory, similarly for the until-concept. We expect that this framework will allow a purely algebraic treatment of equivalence and termination of program schemes. The above construction may be easier performed if the characterization of μf and until(p, f) consists of equations involving only the given functions and predicates, the morphisms to define and morphisms already introduced in this way. By the following lemma partial functions may be replaced by predicates involving no new objects such as A' in 1.2 and 1.6 :

3.1 Lemma: The correspondence $(A \leftarrow A' \xrightarrow{g} B) \overset{\Gamma}{\mapsto} (A\times B \xrightarrow{\chi(\subseteq, g)} 2)$ defines a bijection between (isomorphism-classes of) partial functions from A to B and functional predicates $\gamma : A\times B \longrightarrow 2$, where "functional" means:

$$\begin{array}{ccccccc} A\times B\times B & \xrightarrow{(\pi_o, \pi_1, \pi_o, \pi_2)} & (A\times B)^2 & \xrightarrow{(\gamma^2, \pi_1^2)} & 2^2 \times B^2 & \xrightarrow{\wedge\times =_B} & 2\times 2 \\ \downarrow & & & = & & & \downarrow \Rightarrow \\ 1 & & & \xrightarrow[\text{true}]{\qquad\qquad\qquad} & & & 2 \end{array}$$

The inverse bijection is given by

$$\gamma \overset{\Gamma^{-1}}{\longmapsto} (A \xleftarrow[\pi_o]{} A\times B \leftarrow \overset{-1}{\gamma}(\text{true}) \rightarrowtail A\times B \xrightarrow[\pi_1]{} B) .$$

Functionality of γ implies $\overset{-1}{\gamma}(\text{true}) \longrightarrow A$ to be monic.

The conditions (1) - (3) of 1.2 have a "predicate" equivalent :

3.2 Theorem: Given $f : A\times N \longrightarrow 2$, a predicate $\gamma : A\times N \longrightarrow 2$ is the characteristic function of the graph $A' \xrightarrow{(\subseteq, \mu f)} A\times N$ of a partial function $A \leftarrow A' \xrightarrow{\mu f} N$ satisfying (1) - (3) of 1.2 if and only if γ satisfies the following equations:

(1') $\exists_{\pi_o} \gamma = \exists_{\pi_o} f : A \rightarrow 2$ (in Set defined by $a \mapsto 1$ iff $\exists n \in N : f(a, n) = 1$)

(2') $A\times N \xrightarrow{(\gamma, f)} 2\times 2 \xrightarrow{\Rightarrow} 2 \quad = \quad \text{true}_{A\times N}$

(3') $A\times N\times N \xrightarrow{(\pi_o, \pi_1, \pi_o, \pi_2)} (A\times N)\times(A\times N) \xrightarrow{(\gamma\times f, (\pi_1, \pi_3))} (2\times 2)\times(N\times N)$ —

$\xrightarrow{\wedge\times\leq} 2\times 2 \xrightarrow{\Rightarrow} 2 \quad = \quad \text{true}_{A\times N\times N}$.

3.3 Corollary: Given $f : A\times N \longrightarrow 2$, $\gamma : A\times N \longrightarrow 2$ is uniquely determined by (1') - (3') . This follows by combination of the theorem with proposition 2.2 .

Using 2.5 we can characterize the predicate corresponding to until(p, f) :

3.4 Proposition: Given $A \xrightarrow{f} A \xrightarrow{p} 2$, define

$$\hat{\gamma} := A\times N\times A \xrightarrow{(\Gamma(\mu(pf^{\cdot})), f^{\cdot})\times A} 2\times A\times A \xrightarrow{2\times =_A} 2\times 2 \xrightarrow{\wedge} 2 .$$

Then $\gamma := \exists_{(\pi_o, \pi_2)} \hat{\gamma} : A\times A \longrightarrow A$ corresponds to a partial function (see 3.1) $A \leftarrow A' \xrightarrow{\text{until}(p, f)} A$ satisfying 1.7 (ii) and therefore (1) - (4) of 1.6 .

We hope to have established now a good basis for a subsequent algebraic treatment of some problems of syntax and semantics of recursive program schemes in the framework of "functorial semantics of algebraic theories".

REFERENCES

[1] ALAGIĆ, S., Algebraic aspects of ALGOL 68, Comp. and Inform. Science, Techn. Rep. 73 B-5, Univ. of Mass., 1973

[2] ELGOT, C. C., Monadic computation and iterative algebraic theories, IBM research report RC 4564, Yorktown Heights 1973

[3] FREYD, P. J., Aspects of topoi, Bull. Austr. Math. Soc. 7 (1972), 1 - 76

[4] GOGUEN, J. A. jr., On homomorphisms, correctness, termination, unfoldments and equivalence of flow diagram programs, preprint 1973

[5] JOYAL, A., Arithmetic universes, talk at the Oberwolfach Conf. on Category Theory 1973

[6] LAWVERE, F. W., Functorial Semantics of Algebraic Theories, Proc. Nat. Acad. Sci. USA 50 (1963), 869 - 873

[7] VOLGER, H., Logical categories, semantical categories and topoi, preprint 1973

[8] WAGNER, E. G., An algebraic theory of recursive definitions and recursive languages, Proc. 3rd ACM Symp. Th. Comp. (1971), 12 - 23

[9] WAND, M., A concrete approach to abstract recursive definitions, Artif. Intell. Memo No. 262

[10] VAN DE WAUW - DE KINDER GODELIÈVE, Some properties of the natural number object in a topos, Talk at the Oberwolfach Conf. on Category Theory 1973 .

POWER AND INITIAL AUTOMATA IN PSEUDOCLOSED CATEGORIES

Hartmut Ehrig

Hans-Jörg Kreowski

Technical University of Berlin, FB 20 AT/FS
D-1 Berlin 10, GERMANY

In the last years several attempts have been made to unify automata theory by use of monoidal categories (cf. [1-10]). Decomposition has been studied in [4,5] for several types of automata in monoidal categories. For the case of closed categories, including deterministic, partial, fuzzy-set and several kinds of linear and topological automata, problems concerning reduction, minimization and realization are solved in [2,6,7,8,9]. More difficult is the case of automata in pseudoclosed categories - including nondeterministic, relational, stochastic and relational topological automata - which correspond to Kleisli-Machines studied in [3]. In [6,7,10] we constructed $\mathcal{M}$-minimal (cf. 2.2) and R-reduced (cf. 2.1) automata without initial state even in the pseudoclosed case. As a corollary there are several conclusions like uniqueness up to isomorphism of reduction and, regarding the stateobject only, of $\mathcal{M}$-minimization. Moreover $\mathcal{M}$-minimal automata have minimal cardinality of states. Unfortunately the last two results don't remain true in the case of the initial automata in pseudoclosed categories which will be studied in this paper. However, we can extend the construction of $\mathcal{M}$-minimal and R-reduced automata to the inital case getting weaker properties. Moreover we define a scoop minimization, i.e some states can be replaced by equivalent subsets of the remaining states. These results are developed in § 2. The connection between automata in a pseudoclosed and the corresponding closed category leads to the construction and discussion of power automata in § 1. In fact the well-known equivalence between deterministic and nondeterministic initial automata can be generalized to the case of automata in pseudoclosed categories. The free realization of an arbitrary input-output function by an initial automaton is given in § 3. In addition we give the construction of reachable automata and discuss the compatibility with reduction and minimization.

Unfortunately we cannot give proofs and details in this paper but the reader may consult [6,7,10] for a more detailed version of some constructions and previous results.

General assumptions: Let $(\underline{K},\otimes)$ be a monoidal category with countable coproducts which is pseudoclosed relative $\underline{K}'$ in the sense of [6,7,10], i.e. $(\underline{K}',\otimes)$ is a monoidal closed and coreflexive subcategory of $(\underline{K},\otimes)$ with coreflector P and internal hom-functor $\langle,\rangle$ of $\underline{K}'$. Moreover we assume that $\underline{K}$ and $\underline{K}'$ have the same class of objects and that there is a unique $\mathcal{E}$-$\mathcal{M}$-factorization in $\underline{K}'$. For each automaton A = (I,O,S,d,l) in $\underline{K}$ the machine-function $M(A): S \to \langle I^*\otimes I, PO\rangle$ is a $\underline{K}'$-morphism constructed by adjunction, assigning to each state the corresponding input-output-function. The behavior E(A) of A is given by the $\mathcal{E}$-$\mathcal{M}$-factorization of M(A).

§ 1 POWER AUTOMATA

First we give construction and properties of power automata in the non initial case.

1.1 Theorem (Power Automata): Given an automaton A = (I,O,S,d,l) in $\underline{K}$ the power automaton $\hat{P}A=(I,PO,PS,\hat{d},\hat{l})$ of A, with $\hat{d}$ and $\hat{l}$ defined by $v_s \circ \hat{d} = d\circ(v_s\otimes I)$ and $v_o\circ\hat{l} = l\circ(v_s\otimes I)$ respectively, is an automaton in the closed category $\underline{K}'$ and has the following properties:

1) $\hat{P}$ becomes a right adjoint functor of the inclusion from the category of automata in $\underline{K}'$ to those in $\underline{K}$, both with fixed input I and output-object PO resp O.

2) The machine-functions M(A) of A and $M(\hat{P}A)$ of $\hat{P}A$ satisfy

a) $M(\hat{P}A) = \sigma\cdot PM(A)$ and

b) $M(A) = M(\hat{P}A)\circ i_s$ with i_s unit of the adjunction $J \dashv P$.

The morphism $\sigma : P(\langle I^*\otimes I, PO\rangle) \longrightarrow \langle I^*\otimes I, PO\rangle$ in a) is a canonical operation on $\langle I^*\otimes I, PO\rangle$ corresponding to the union of relations for example.

3) The following closure properties for the behavior are valid:

a) $E(A) \subseteq E(\hat{P}A) = E(\hat{P}\hat{P}A)$ and

b) $E(A) \subseteq E(A')$ implies $E(\hat{P}A) \subseteq E(\hat{P}A')$.

By 2a) A is strong $\mathcal{M}$-minimal in the sense of [7] if and only if $\hat{P}A$ is $\mathcal{M}$-minimal and the power automaton $\hat{P}A$ realizes the closure of E(A) with respect to the union σ. The closure of E(A), however, with respect to multiplication with input strings, is given by a subautomaton of $\hat{P}A$ called kernel automaton $\hat{K}A$ which is the smallest automaton in $\underline{K}'$ simulating A(cf. [10]).

The construction of power automata can be extended to the initial case such that we get the equivalence mentioned in the introduction.

1.2 Definition: An initial automaton A=(I,O,S,d,l,a) in a pseudoclosed category $\underline{K}$ consists of objects I,O,S and $\underline{K}$-morphisms $d:I\otimes S \longrightarrow S$, $l:I\otimes S \longrightarrow O$ and $a:E \longrightarrow S$, E unit object in $(\underline{K},\otimes)$. The behavior $E^{\bullet}(A)$ of an initial automaton A is given by

$$E^{\bullet}(A)=(I^{*}\otimes I \xrightarrow{\sim} E\otimes I^{*}\otimes I \xrightarrow{a\otimes I^{*}\otimes I} S\otimes I^{*}\otimes I \xrightarrow{d!\otimes I} S\otimes I \xrightarrow{l} O).$$

Given initial automata A and A' (I and O fixed) a $\underline{K}$-morphism $f:S \longrightarrow S'$ is called automaton morphism $f:A \longrightarrow A'$ if $f\circ d=d'\circ(f\otimes I)$, $l=l'\circ(f\otimes I)$ and $f\circ a=a'$. Hence we get the category $\underline{K}$-$\underline{Aut}^{\bullet}(I,O)$ of initial automata in $\underline{K}$. Moreover each morphism $f:A \longrightarrow A'$ in $\underline{K}$-$\underline{Aut}^{\bullet}(I,O)$ implies $E^{\bullet}(A)=E^{\bullet}(A')$ such that the behavior defined above can be extended to a functor $E^{\bullet}:\underline{K}$-$\underline{Aut}^{\bullet}(I,O) \longrightarrow \underline{B}^{\bullet}$, where $\underline{B}^{\bullet}$ is the discrete category of all $\underline{K}$-morphisms $b:I^{*}\otimes I \longrightarrow O$. Restricting the automaton morphisms to be $\underline{K}'$-morphisms we write $\underline{K}$-$\underline{K}'$-$\underline{Aut}^{\bullet}(I,O)$ for the corresponding category.

In analogy to theorem 1.1 we get:

1.3 Theorem (Initial Power Automata): Given an initial automaton A=(I,O,S,d,l,a) in $\underline{K}$ the power automaton $\hat{P}A=(I,PO,PS,\hat{d},\hat{l},\hat{a})$ of A with d and l as given in 1.1 and $a:E \longrightarrow PS$ defined by $v_S\circ\hat{a}=a$ is an automaton in $\underline{K}'$ and has exactly the property given in 1.1,1) with $\underline{Aut}$ replaced by $\underline{Aut}^{\bullet}$. Moreover $\hat{P}A$ and A are equivalent, i.e. $E^{\bullet}(\hat{P}A) = E^{\bullet}(A)$.

§ 2 REDUCTION AND MINIMIZATION

The constructions of $\mathcal{M}$-minimal and R-reduced automata given in [7] § 2 can be carried over to the initial case. In addition we can give an interesting scoop construction replacing some states by equivalent subsets of the remaining states.

2.1 Theorem (R-reduced Automata): An automaton A is called R-reduced, if each reduction $f:A \longrightarrow A'$, i.e. $f\in\mathcal{E}$, is already an isomorphism. The subcategory of $\underline{K}$-$\underline{K}'$-$\underline{Aut}^{\bullet}(I,O)$ defined by all R-reduced automata A is $\mathcal{E}$-reflexive such that the reflexion morphisms $u(A):A \longrightarrow R(A)$ are reductions preserving the behavior.

As in the non initial case we have to assume that $\underline{K}'$ is cocomplete and cowellpowered for this theorem and the solution is given by a cofibre product.

<u>2.2 Theorem ($\mathcal{M}$-minimal Automata)</u>: An automaton is called $\mathcal{M}$-minimal if we have $M(A) \in \mathcal{M}$ for the machine function of A. Let $\underline{K}$-$\underline{K}'$-$\underline{Aut}^{\bullet}_{w}(I,O)$ be the category of initial automata with weak morphisms $f:A \longrightarrow A'$, i.e. $f \in \underline{K}'$ satisfies $M(A) = M(A') \circ f$ and $f \circ a = a'$. Then the subcategory of all $\mathcal{M}$-minimal automata is $\mathcal{E}$-reflexive, provided that all morphisms in $\mathcal{E}$ have coretractions in $\underline{K}$. Especially for each A there is an equivalent $\mathcal{M}$-minimal automaton A'.

Now we will be concerned with the scoop construction:

<u>2.3 Definition:</u> Given an initial automaton $A=(I,O,S,d,l,a)$ in $\underline{K}$ a pair $(m:S' \longrightarrow S,\ n:S \longrightarrow PS')$ with $m \in \mathcal{M}$ and $n \in \underline{K}'$ is called scoop of A, if $M(A)=M(\hat{P}A) \circ Pm \circ n$. A is called scoop-minimal if each scoop (m,n) of A is trivial, i.e. m is an isomorphism.

<u>Remark:</u> Each scoop defines a smaller automaton $A(m,n)$ with states S' which is equivalent to A. Moreover scoops are closed under composition and hence for each "finite" automaton A there is a scoop (m,n) of A such that $A(m,n)$ is scoop-minimal.

<u>2.4 Motivation and Construction:</u> In order to get a scoop for a given automaton $A=(I,O,S,d,l,a)$ we first construct the pullback L with projections $p_1:L \longrightarrow S, p_2:L \longrightarrow PS$ of the machine-functions $M(A)$ and $M(\hat{P}A)$. In all our examples L is the set of all pairs $(s_1,S_1) \in S \times PS$ such that s_1 is equivalent to S_1, i.e. $M(A)(s_1)=M(\hat{P}A)(S_1)$. Clearly s_1 can be replaced by S_1 in order to get a minimization of A, but this only makes sense in the case $S_1 \neq \{s_1\}$. Now we give a construction for the set S^* of all such elements s_1 which can be replaced by subsets S_1. By theorem 1.1 we have $M(\hat{P}A) \circ i_s = M(A)$ such that there is a unique morphism $u:S \longrightarrow L$ satisfying $p_1 \circ u = 1_S$ and $p_2 \circ u = i_s$. Assuming $\underline{K}'$ to be <u>Sets</u> or any well-pointed topos the coretraction $u:S \longrightarrow L$ is in fact a coproduct injection with complement $\bar{u}:(L-S) \longrightarrow L$. Thus we get the desired S^* as the $\mathcal{E}$-$\mathcal{M}$-factorization $(L-S) \xrightarrow{e^*} S^* \xrightarrow{m^*} S$ of $p_1\bar{u}$.

<u>2.5 Theorem (Scoop-Construction):</u> For each automaton A as given in 2.4 there is a scoop $(m:S' \longrightarrow S, n:S \longrightarrow PS')$ of A given as follows: The $\underline{K}'$-morphism $q:=(S^* \xrightarrow{c} (L-S) \xrightarrow{\bar{u}} L \xrightarrow{p_2} PS)$, with c being a coretraction of $e^* \in \mathcal{E}$, has a factorization $q=S^* \xrightarrow{q'} PS' \xrightarrow{Pm} PS$ where $m:S' \longrightarrow S$ is the intersection of all $\mathcal{M}$-subobjects admitting such a factorization of q and containing $S-S^*$. $n:S \longrightarrow PS'$ is the unique morphism induced by $q':S^* \longrightarrow PS'$ and $Pm' \circ i_{S-S^*}:(S-S^*) \longrightarrow PS'$ where m' is the inclusion $m':(S-S^*) \longrightarrow S'$.

Moreover we have that the resulting automaton A(m,n)(cf.2.3) is scoop-minimal if S'equals S-S*. In fact in this case we have the stronger condition that S-S* is pullback of the machine-functions M(A(m,n)) and M($\hat{P}$A(m,n)), i.e. no state s_1 is equivalent to a subset $S_1 \neq \{s_1\}$ in our examples.

Remark: The theorem is applicable to nondeterministic, relational and stochastic automata. The third part of 2.5 is a generalization of a minimization theorem for nondeterministic automata given in [11].

§ 3 REALIZATION AND REACHABILITY

In this paragraph we give the constructions of free realization and reachable automata Moreover we discuss compatibility with reduction and minimization.

3.1 Theorem (Free Realization): For each input-output function $b:I^* \otimes I \longrightarrow O$ in $\underline{K}$ the automaton $F(b)=(I,O,I^*,\mu,b,i_o)$, called free realization, with $\mu:I^*\otimes I \longrightarrow I^*$ restriction of the monoid multiplication and $i_o:E \longrightarrow I^*$ unit of I^*, realizes b, i.e. $E^{\bullet}F(b)=b$, and F becomes a left adjoint functor of $E^{\bullet}:\underline{K\text{-}Aut}^{\bullet}(I,O) \longrightarrow \underline{B}^{\bullet}$

3.2 Definition: An automaton A=(I,O,S,d,l,a) is called reachable if the morphism $d!\circ(a\otimes I^*):E\otimes I^* \longrightarrow S$ belongs to a given class $\tilde{\mathcal{E}}$ of not necessarily epimorphisms in $\underline{K}$. We presume to have an $\tilde{\mathcal{E}}$-$\mathcal{M}$-factorization in $\underline{K}$ compatible with $\mathcal{E}$ in $\underline{K}'$.

3.3 Theorem (Reachability): For each automaton A=(I,O,S,d,l,a) there is an equivalent reachable automaton $C(A)=(I,O,B,d_B,l_B,a_B)$ defined in the following way: Let $E\otimes I^* \xrightarrow{\tilde{e}} B \xrightarrow{\tilde{m}} S$ be an $\tilde{\mathcal{E}}$-$\mathcal{M}$-factorization of $d!\circ(a\otimes I^*)$ which is a right action-morphism, i.e. compatible with m,i_o and d!,a. Hence there is a unique diagonal-morphism $p_B:B\otimes I^* \longrightarrow B$ making $\tilde{e}$ and $\tilde{m}$ right action morphisms. Now take $d_B=p_B\circ(B\otimes i_1)$, $l_B=l\circ(\tilde{m}\otimes I)$ and $a_B=e\circ i_o$.

Moreover we have:

a) C becomes a right adjoint functor of the inclusion of reachable automata.

b) Reachability is compatible with reduction and minimization in the following sense: A R-reduced implies C(A) R-reduced and R (A) is reachable if this is true for A. Moreover the composite functors $R_{\tilde{\mathcal{E}}}\circ C$ and $C_R\circ R$ are natural equivalent where $R_{\tilde{\mathcal{E}}}$ and C_R are restrictions of R and C respectively. If A is $\mathcal{M}$-minimal then also C(A). For reachable A with $\mathcal{M}$-minimization A' there is a reachable $\mathcal{M}$-minimal automaton equivalent to A.

<u>3.4 Theorem (Behavior and $\mathcal{M}$-minimal Realization)</u>

1) Given an input-output-function $b: I^* \otimes I \longrightarrow O$ in $\underline{K}$ the $\mathcal{M}$-minimization $\widetilde{M}(b)$ of the free realization $F(b)$ of b coincides with the R-reduction $R \circ F(b)$ up to isomorphism and both $F(b)$ and $\widetilde{M}(b) \cong R \circ F(b)$ are in fact $\widetilde{\mathcal{E}}$-reachable, since their next-state-functions belong to $\underline{K}'$, i.e. "deterministic". $\widetilde{M}(b)$ is called $\mathcal{M}$-minimal realization of b with state object $\widetilde{S}(b) \cong E\widetilde{M}(b) \cong EF(b)$.

2) Moreover for an arbitrary realization A of b, i.e. $E^{\bullet}(A) = b$, we have $E\,\widetilde{M}(b) \subseteq E\,\hat{P}(A)$ where $\hat{P}(A)$ is the power automaton of A.

3) Given a full subcategory $\underline{L}$ of $\underline{K}'$ - closed under $\mathcal{E}$-homomorphic images, $\mathcal{M}$-subobjects and under the functor P which is assumed to preserve $\mathcal{E}$ - we have the following characterization: An input-output-function $b: I^* \otimes I \longrightarrow O$ in $\underline{K}$ is realizable by an automaton with state object in $\underline{L}$ iff $\widetilde{S}(b)$, i.e. the states of the behavior b, belongs to $\underline{L}$.

<u>Remarks</u>: A minimal realization construction similar to that of initial automata in closed monoidal categories (cf. [6,8,9]) does not exist in the pseudoclosed case. The best general approximation seems to be given by the scoop-construction applied to the $\mathcal{M}$-minimal realization constructed in 1) of theorem 3.4.

Statement 2) in our applications implies that for each function $b' \in E\widetilde{M}(b)$ there is a set of input-output functions in the behavior E(A) of A such that b' is the union of this set (cf. 1.1,2). For the relational case we get the following inequality for the cardinality card(S) of states of an arbitrary realization A of b:

$$\mathrm{card}(S) \geq \log_2(\mathrm{card}(\widetilde{S}(B))$$

Taking $\underline{L}$ to be finite sets statement 3) is applicable to the nondeterministic and relational case characterizing input-output-functions realizable by finite automata.

REFERENCES

[1] ARBIB, M.A. - MANES, E.G., Machines in Category: an Expository Introduction, Siam Review, to appear

[2] -, Foundations of System Theory I, Automatica, to appear

[3] -, Kleisli Machines, to appear

[4] BUDACH, L. - HOEHNKE, H.-J., Über eine einheitliche Begründung der Automatentheorie, Seminarbericht 1. Teil, HU-Berlin, 1969/70

[5] -, Automaten u. Funktoren, Berlin 73/74, to appear

[6] EHRIG, H. - KIERMEIER, K.D. - KREOWSKI, H.-J., KÜHNEL, W., Systematisierung der Automatentheorie, Seminarbericht 72/73 TU Berlin FB 20 Bericht 73-08, to be published in English by Teubner

[7] EHRIG, H. - KREOWSKI, H.-J., Systematic Approach of Reduction and Minimization in Automata and System Theory, TU Berlin 1973, FB 20 Bericht 73-16, submitted for JCSS

[8] GOGUEN, J.A., Systems and Minimal Realization, Proc. of the 1971 IEEE Conf. on Decision and Control, Miami Beach 1971, 42-46

[9] -, Discrete-Time-Machines in Closed Monoidal Categories, I. Quarterly Report no. 30, Inst.f.Comp.Res., University of Chicago (1971). Condensed version in: Bull. AMS 78 (1972), 777-783

[10] KREOWSKI, H.-J., Automaten in pseudoabgeschlossenen Kategorien, Diplomarbeit TU-Berlin 1974

[11] STARKE, P.H., Abstrakte Automaten, VEB-Verlag, Berlin 1969 (Translation: Abstract Automata, Elsevier/North Holland, Amsterdam 1972)

SEMANTICS OF COMPUTATION[1]

J. A. Goguen
Computer Science

University of California
Los Angeles, CA 90024 U.S.A.

0. INTRODUCTION

There has been a remarkable convergence of the problems of practical computer science and system theory with the methods of categorical algebra. One meeting ground for these concerns is realization theory, which considers methods for finding systems, preferably "minimal" in some sense, to perform some task (i.e., to "realize" some "behavior"). The problems of classical control thoery are of this type: a feedback control system is sought for some "plant," or whatever. Computer programming is also a realization problem, in which computer code is written to perform some given calculation. Because programming is done largely by humans, "bugs," i.e. errors, are one of the main sources of difficulty in using computers, and thus, one of the main areas of attention in computer science. "Program correctness" and "structured programming" are methodologies to achieve "software reliability."

The "semantics" of a system is its behavior. From a broad point of view, semantics and realization are aspects of the same situation: semantics is the problem of system analysis; while realization is the problem of system synthesis.

---

[1]This research was supported (in part) by the National Science Foundation, Grant No. GJ.33007X.

This paper is a brief and hopefully understandable survey of various aspects of semantics, particularly in relation to computational methods; the primary objective is to set forth an algebraic point of view. Sometimes details must be sought elsewhere; on the other hand, some results are announced for the first time.

I wish to express my thanks to all those with whom I have interacted in this area, and in particular to: R. Burstall, and M. Arbib; to my collaborators and the "ADJ" project, Drs. J. W. Thatcher, E. G. Wagner, and J. B. Wright; and to my students L. Carlson and S. Ginali.

1. INITIAL ALGEBRA SEMANTICS

Algebraic structures have much to do with computation. For a start, "operations to be performed" are the basis for any procedure. One must first distinguish between: the symbols which denote the operations, as part of the instructions for the computational procedure; and the functions as actually executed. This is the distinction between syntax and semantics; or between program and processor.

Let us consider first the case of unconditional, i.e. non-branching computations, such as

$$(1)\quad ((2\cdot(1+X_0)) + ((-X_1)\cdot(X_2\uparrow 2)))$$

($\uparrow$ means "to the power"). This involves a set $\Omega_0 = \{1, 2, X_0, X_1, X_2\}$ of given values ("constants"), a set $\Omega_2 = \{+, \cdot, \uparrow\}$ of binary operations, and a set $\Omega_1 = \{-\}$ of unary operations. The general framework lets an "operator domain" Ω be a sequence $\Omega_0, \Omega_1, \Omega_2, \ldots$ of sets, and defines the set T_Ω of Ω-terms to be the least subset of $(\bigcup_{n=0}^{\infty}\Omega_n \cup \{(\ ,\ )\})^*$ such that:

(t0) $\Omega_0 \subseteq T_\Omega$; and

(t1) for $n \geq 1$, $\sigma \in \Omega_n$, and $t_1, \ldots t_n \in T_\Omega$, $\sigma(t_1 \ldots t_n) \in T_\Omega$.

T_Ω is the set of all "non-branching" programs in Ω (using a slightly different notation than the "infix" of (1) above); it plays the role of "syntax."

A possible semantics or <u>processor</u> for these programs is an <u>Ω-algebra</u>, that is, a set A, called the <u>carrier</u>, and a function σ_A: $A^n \to A$ for each $\sigma\varepsilon\Omega_n$; for n = 0, this means a constant $\sigma_A \varepsilon A$. (We may omit the subscript.) For example, using the Ω suggested by (1), let A = $\{+, -\} \times \{0,1\}^{36}$, let 1_A = +0 ... 01, let $-_A$: $A \to A$ reverse signs, let $+_A$, $\bullet_A$, $\uparrow_A$ be some particular (unspecified) 36-bit approximations to the operations these symbols usually denote, and finally, let $(X_i)_A$, for i = 0, 1, 2, be some (particular but unspecified) available constants (possibly read from hard wired external registers). <u>Execution</u> of programs on this processor should be a function h: $T_\Omega \to A$.

In fact, T_Ω is also an Ω-algebra, with $\sigma_T : T^n_\Omega \to T_\Omega$ (for $\sigma\varepsilon\Omega$, $n \geq 1$), defined to send $(t_1, \ldots, t_n)$ to $\sigma(t_1 \ldots t_n)$; and for $\sigma\varepsilon\Omega_0$, take δ_T to be just σ itself. Moreover, h should be an <u>Ω-homomorphism</u>, in the sense that:

(h0) $h(\sigma_T) = \sigma_A$ for $\sigma\varepsilon\Omega_0$; and

(h1) for $n \geq 1$, $\sigma\varepsilon\Omega_n$, and $t_1, \ldots, t_n \varepsilon T_\Omega$,

$$h(\sigma_T(t_1, \ldots, t_n)) = \sigma_A (h(t_1), \ldots, h(t_n)).$$

(This definition evidently generalizes to functions h:T→A between any Ω-algebras.) For the case at hand, the conditions (h0), (h1) say that h "correctly executes" each individual instruction σ in Ω. Actually, there is one and only one way of executing programs if each instruction is correctly executed, as the following result asserts.

<u>Proposition 1</u>. T_Ω is an initial object in the category of Ω-algebras.

We take this as prototypical of a wide range of situations in computational semantics.

Note that elements of T_Ω can be represented as trees. For example, (1) can also be written:

(2) 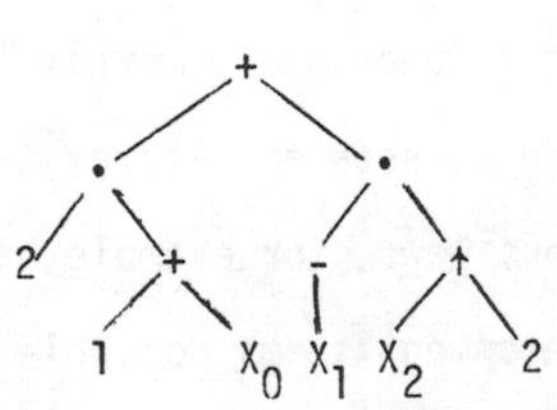

Compilers produce representations of the form (2), because it indicates the order operations can be performed (for example, we must fetch the numbers 1 and X_0 before we can compute their sum).

A tree automaton is an Ω-algebra A with a designated subset $F \subseteq A$ of final states. The set recognized by <A,F> is $h^{-1}F \subseteq T_\Omega$. This generalizes the usual concepts of automaton and regular set, which are the special case with $\Omega_0 = \{0\}$, $\Omega_1 = X$ and $\Omega_n = \emptyset$ for $n \geq 2$. Then T_Ω is (isomorphic to) X^* (the set of all strings of elements from x), and $S \subseteq X^*$ is regular iff $S = h^{-1}F$ for $F \subseteq$ some finite Ω-algebra A. See Thatcher [1], Thatcher-Wright [2], Eilenberg-Wright [3].

Definition. An Ω-tree is an element of an initial Ω-algebra.

For example, the "Vienna definition language" speaks of the "characteristic set" of a tree as its applicable "compound selectors"; this can be generalized to the set of all root-to-frontier paths and can be obtained by applying the unique function h: $T_\Omega \to A = P(\cup_n \Omega_n)^*$, where P denotes "power set of" and A is given Ω-algebra structure by defining $\sigma_A: A^n \to A$ (for $\sigma \in \Omega_n$) to send $\langle P_1, \ldots, P_n \rangle$ to the set $\{\sigma p_1 \ldots p_n | p_i \in P_i,\ i=1, \ldots, n\}$, and for $\sigma \in \Omega_n$ letting σ_A be $\{\sigma\}$. I.e., this function on trees is defined by Proposition 1; similarly one can define functions such as frontier, depth, and root.

Such definitions really generalize ordinary induction, which is the special case with $\Omega = \{0\}, \{\sigma\}, \emptyset, \emptyset, \ldots$ (see, e.g. Maclane-Birkhoff [5]: the non-negative integers $\omega = \{0, 1, 2, \ldots\}$ are an initial algebra for this Ω, with $\sigma(n) = n+1$). In fact, Proposition 1 is really a generalized postulate of induction which can often replace (as a challenge, "always replace") explicit appeals to inductive or recursive arguments.

Consider an equational variety of Ω-algebras (i.e., a category of $\underline{T}$-algebras for a (Lawvere) theory $\underline{T}$). Then there is still an initial object, but it is no longer "anarchic", merely "free"; its elements can be represented as equivalence classes of Ω-trees, and the processors may now be taken to satisfy various laws, for example, a commutative law for +. The full theoretical development seems possible only in the free case, but the categorical machinery helps. See Eilenberg-Wright [3].

We now consider "many-sorted" algebras. Modern programming languages, such as ALGOL 60, have several "types" or "sorts" of variables, such as integer, boolean, real, and vector. There are also interesting connections with context free languages and recognizable sets. Maibaum [6], Turner [7], and Wand [8] have obtained an extension of the Chomsky hierarchy to include "indexed languages." Here, we show how "Knuth semantics" [9] arises in the present framework.

First, the basic definitions: An <u>S-sorted operator domain</u> Ω (for S a set of "sorts") is a family $\Omega_{w,s}$ of sets, indexed by pairs $w,s \in S^* x S$; $\Omega_{w,s}$ is the set of symbols for operators which take arguments of sorts $s_1, \ldots, s_n = w \in S^*$ and return value of sort $s \in S$. Then an <u>Ω-algebra</u> A is a family $<A_s>_{s \in S}$ of sets (one for the elements of each sort), and a function $\sigma_A: A^w \to A_s$ for each $\sigma \in \Omega_{w,s}$ where A^w means $A^{s_1} x \ldots x A^{s_n}$ (for $w = s_1 \ldots s_n$; A^{Λ} denotes some one point set). An Ω-homomorphism h: $A \to B$ is a family of functions $<h_s : A_s \to B_s>_{s \in S}$ such that for each $\sigma \in \Omega_{w,s}$

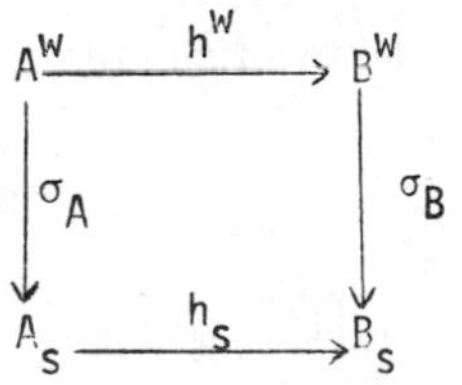

commutes (where h^w means $h_{s_1} x \ldots x h_{s_n}$ for $w = s_1 \ldots s_n$). Once again, the evident recursive definition yields an initial Ω-algebra, denoted T_Ω. See ADJ [4] for examples and more details.

Now let us consider a context free grammar G to be a triple $<V_N, V_T, P>$ with $P \subseteq V_N X V_N^+ \cup V_N X V_T$ (where X^+ means $X^* - \{\Lambda\}$); here V_N is the non-terminal vocabulary, and V_T the terminals. This determines a V_N-sorted operator domain, also denoted G, by setting $G_{w,s} = \{s \mid <s,w> \in P,\ w \in V_N^+\}$, $G_{\Lambda,s} = \{a \mid <s,a> \in P\}$. Then T_G is the set of all derivation trees in G of terminal strings (and the carrier $(T_G)_S$ is those starting from $S \in V_N$).

If A is G-algebra, there is a unique G-homomorphism $T_G \to A$. This is the setting for Knuth's [9] semantics!!! For example, let G be part of a grammar for ALGOL 60 [10] containing the production

(3) <arithmetic expression>→<if><Boolean expression><then><simple arithmetic expression><else><arithmetic expression>,

and A be a G algebra with $A_{\text{<arithmetic expression>}} = A_{\text{<simple arithmetic expression>}}$ rather like our earlier A (now denoted A_a), with $A_{\text{<if>}} = A_{\text{<else>}} = A_{\text{<then>}}$ = some one pointed set, (which we shall ignore), and with $A_{\text{<Boolean expression>}} = \{0,1\}$. Then A should contain a function "for evaluating arithmetic expressions."

(4) $e: \{0,1\} X A_a X A_a \to A_a$

such that $e(i, x_1, x_0) = x_i$ for $i \in \{0,1\}$. Similarly, the production

<term>→<term><X><factor>

in G, will have in A the evaluating function

$X: A_a X A_a \to A_a$.

We remark that Scott type semantics [11], [12] can also be done by initial object methods; as in earlier cases, the difficulty lies more in the construction of the object than in the semantics itself, which follows directly from initiality. One of the additional factors which must be confronted is continuity, which we discuss later in this paper. Finally, it is worth noting that initiality gives an elegant definition of substitution, from which such properties as associatively follow easily (see ADJ [4]).

2. FLOW DIAGRAM PROGRAMS

Consider the following simple program P

```
(5)          READ X,Y
             Z:=0
             If Y=0
               PRINT Z
               END
       (L)   If Y>0
               Z:=2Z+X
               Y:=Y-1
               GOTO (L)
```

whose "semantics" could be the following "flow chart,"

(6)

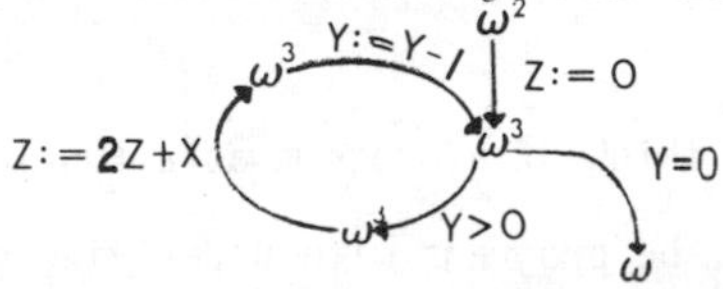

in which nodes are labelled with sets ($\omega=\{0,1,2,\dots\}$, $\omega^2=\{\langle x,y\rangle \mid x,y\varepsilon\omega\}$, $\omega^3=\{\langle x,y,z\rangle \mid x,y,z\varepsilon\omega\}$), and edges with partial functions. For example, Y:=Y-1 sends $\langle x,y,z\rangle$ to $\langle x,y-1,z\rangle$, and is defined iff $y\geq 1$; Y>0 is the partial subfunction of the identity function $\omega^3\to\omega^3$ defined only for those $\langle x,y,z\rangle$ with $y>0$.

What function does P compute? For what inputs does it halt? Is there a simpler program computing the same function? These are fundamental issues in program semantics. Our discussion follows Goguen[13], inspired by Burstall[14].

We shall assume programs given the form (6); more precisely, a flow diagram program is a graph G with nodes labelled by sets and edges by partial functions, so that if $e:v_0\to v_1$ is an edge, then its label is $P(e):P(v_0)\to P(v_1)$. Equivalently, such a program is a graph morphism $P:G\to\underline{\underline{Pfn}}$, where $\underline{\underline{Pfn}}$ denotes the (underlying graph of the) category of sets with partial functions. The mathematical semantics of such a program, from a node v_0 as input to v_1 as output, is the relation

(7) $\qquad \cup\{\overline{P}(f) \mid f:v_0\to v_1\text{, a path in } G\}$,

denoted $P(v_0, v_1)$. If $f=e_1\dots e_n$ is a path, then $\overline{P}(f)$ denotes the composition $P(e_1)\dots P(e_n)$; union in (7) refers to functions as sets of ordered pairs.

The union of functions need not be a function, but actually, each $P(v_0, v_1)$ is a (partial) function provided that all the functions on the edges leaving each fixed node, have disjoint domains of definition. This holds for (6), which in fact computes $(2y-1)\cdot x$. We shall prove this with program homomorphisms. Let $G^{\circledast}$ denote the category of all paths in G; i.e., $G^{\circledast}(v_0,v_1)=$ all paths $v_0 \to v_1$ in G; and composition is juxtaposition. Let $i_G: G \to G^{\circledast}$ be the inclusion graph morphism.

Proposition 2. G^{*} is the free category generated by G; i.e. for any category $\underline{\underline{C}}$ and graph morphism $P: G \to \underline{\underline{C}}$, there is a unique functor $\overline{P}: G^{\circledast} \to \underline{\underline{C}}$ such that $i_G \overline{P} = P$.

In particular, we can think of a program as a functor $\overline{P}: G^{\circledast} \to \underline{\underline{\mathrm{Pfn}}}$.

Definition. Let P_0, P_1 be programs with underlying graphs G_0, G_1. Then a homomorphism $P_0 \to P_1$ is a graph morphism $F: G_0 \to G_1^{\circledast}$ and a natural transformation $n: P_0 \Rightarrow \overline{P}_1 \overline{F}$. $\langle F, n\rangle$ is a simulation iff each value n_v of n is an inclusion, and a projection iff $\overline{F}$ and each of n_v are total surjective.

Using Proposition 2, we can show that given F, a family $n_v: P_0 v \to P_1 F_v$ is natural iff the naturality condition holds for each edge in G_0. The following justifies "Floyd's method"[15].

Proposition 3. If $\langle F, n\rangle: P_0 \to P$ is a simulation, then $P_0(u,v) \subseteq P(Fu, Fv)$.

For fixed $x, y \in \omega$, let P_0 be the following program

(8)

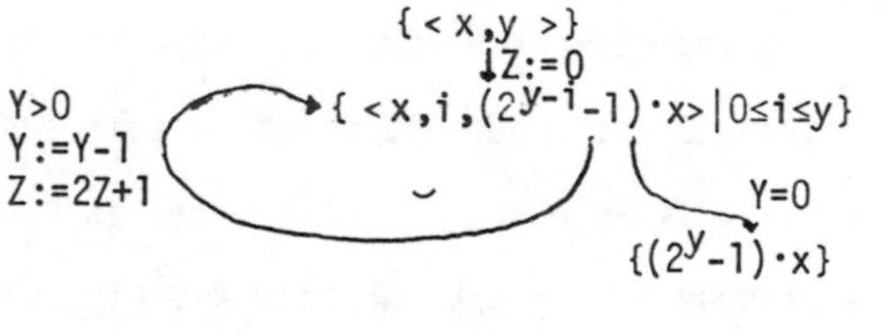

Note that its underliying graph is a little simpler than (6), while one of its operations is more complex, and all its sets are subsets of those of (6). It must be checked that the functions go into their targets. Define $F: G_0 \to G^{\circledast}$ the obvious way (sending the "loop" edge in (8) to the "loop path" in (6)), and let each n_v be inclusion. Then naturality is obvious, and we conclude $P_0(u,v) \subseteq P(u,v)$. But the semantics of P_0 must be $\langle x,y\rangle \mapsto (2^y-1)\cdot x$, or else the empty function. Let us assume that P_0 terminates (we prove this soon).

Then the semantics of P must also be $< x,y > \mapsto (2^y - 1)\cdot x$; for this relation is actually a function, and Proposition 3 gives one value for each input.

Let v, v' be nodes of P; then we say P terminates at v from v' provided P(v,v') is total. The following sometimes allows much simpler proofs of termination.

Proposition 4. If $P_0 \to P_1$ is a projection, then for all nodes v, v' in P_1, P terminates at $F^{-1}v$ from $F^{-1}v'$ whenever P_1 terminates at v from v'.

We apply this to the program P_0 of (8): let P_1 be the program

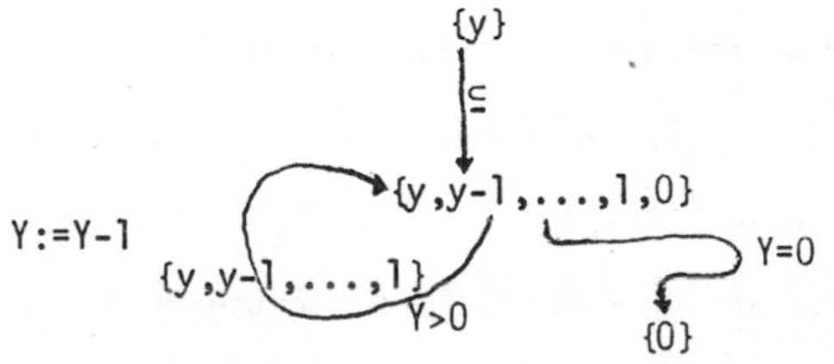

and let $< F;n >$ be the evident projection (to the Y component). Then P_1 certainly terminates; therefore P_0 does too. The reader might compare this with Manna's[16] awkward unsatisfiability approach.

Programs with their homomorphisms form a category Prog. Goguen[13] has shown the pointed programs whose underlying graphs are trees forms a coreflective subcategory, with coreflector the unfoldment functor. This subcategory if cut down to inclusion morphisms, is close to Scott's lattice of flow diagrams[11]. It enjoys the following important continuity property: closure under (non-void) countable filtered colimits. This seems to be the correct generalization of the continuity conditions of "fixed point semantics" and related areas, and leads to an initial algebra semantics for the Scott theory.

If we had more time, we would go into the questions of program schemes and recursion semantics, which require V-categories and V-theories, in some sense. Burstall-Thatcher[17] raises some of these issues, and Burstall (unpublished notes) created "vel-categories." Goguen[13] mentions V-theories, and Wagner (unpublished notes) has considered ordered theories and their completions. We believe it is promising to take V to be: (countably) complete lattices with morphisms preserving suprema of (non-void countable) directed sets; or (countably) cofiltered-cocomplete categories.

3. NON-LINEAR SYSTEMS

We can think of a machine M as defining a (generally non-linear) computational method. In particular, the 6-tuple model $M=\langle X,S,Y,\delta,\beta,\sigma\rangle$ (where X,S,Y are the input, state, output sets; $\delta:S\times X\to S$, $\beta:S\to Y$ are the next state, output functions; and $\sigma\in S$ is the initial state) has as semantics a (external) input-output behavior function $E(M):X^*\to Y$, defined by $E(M)(w)=\beta(\overline{\delta}(\sigma,w))$, where $\overline{\delta}$ is defined recursively by $\overline{\delta}(s,\Lambda)=s$ and $\overline{\delta}(s,x_1\ldots x_n)=\delta(\overline{\delta}(s,x_1\ldots x_{n-1}),x_n)$.

Machines, with morphisms triples $\langle a,b,c\rangle$ of functions on inputs, states, outputs such that δ,β,σ are preserved, form a category Mach, which is in fact the category of $\{X,S,Y\}$-sorted Ω-algebras. Behaviors also form a category Beh with morphisms $B\to B'$ pairs $\langle a,c\rangle$ of functions on inputs, outputs such that $B'a^*=bB$. Then semantics is a functor $E:$Mach$\to$Beh. The realization problem requests a functor N (after Nerode who gave the ur-construction) the other way, such that $E(N(B))=B$. A machine is minimal iff reduced and reachable i.e., iff all states are inequivalent and reached ($s,s'\in S$ are equivalent iff $\overline{\delta}(s,w)=\overline{\delta}(s',w)$, all $w\in X^*$; and $s\in S$ is reached iff there is some $w\in X^*$ such that $\overline{\delta}(\sigma,w)=s$). For S finite, M is minimal iff no M' with $E(M)=E(M')$ has fewer states; but for infinite state machines, such a condition won't do.

The functor $R:$Mach$\to$Mach taking each M to its reachable submachine (state set $\overline{\delta}(\sigma,X^*)\subseteq S$) is determined (up to isomorphism of state sets) as a right adjoint to the inclusion of reachables in Mach such that $ER=E$ (the adjunction was first pointed out to me by L. Carlson in 1968-69).

The crucial relationship between the functors E,R,N is relative adjunction, that is, a natural isomorphism

(9) $$\mathrm{Beh}(E(M),B)\cong\mathrm{Mach}(R(M),N(B))$$

which uniquely determines N (up to isomorphism of states, given EN= the identity on Beh, and this provides an alternative definition of minimality via universal arrows. Such considerations seem to have great generality, for example for machines in suitable monoidal categories, and for algebra (i.e. tree) machines; but not for stochastic or for non-deterministic machines, which seem to be poorly endowed with pleasing theoretical results.

The special case of (9) in which R is the identity, for example where Mach contains only reachable machines, leads to the study of right-adjoint-right-inverse situations, called minimal realization situations (MRS's) by Goguen[18]. and also studied by Ehrig-Kreowski[19]. Such an approach applies to the syntax/semantics relationship of Lawrere[20], as well as to initial factorizations (ADJ[4]), and many other aspects of algebraic theories.

Sample Result 5. Functors $E:\underline{\underline{M}}\to\underline{\underline{B}}$, $N:\underline{\underline{B}}\to\underline{\underline{M}}$ form a MRS iff there is a natural transformation $n:\underline{\underline{M}}\Rightarrow EN$ such that $nE=E$ and $Nn=N$, and E is surjective on objects.

In a sense, such a theory is too general, because it fails to yield powerful existence theorems for machines. However, such theorems exist for machines in "suitable" categories, and have the additional glory of unifying discrete machines (as above) with linear and species of non-linear control systems. This seems to have been first done by Goguen[21], who also singled out the cases of topological and affine machines, which live (respectively) in the categories Kell of Kelley (i.e., compactly generated Hausdorff) spaces, and $\underline{\underline{Aff}}_k$ of K-modules (for K a ring with unit) with K-affire morphisms (K-linear plus constant). Each category has a "product" which accounts for the non-linear character of its machines. Cartesian product in Kell gives a transition function $\delta:X\times S\to S$ in Kell iff continuous merely on each compact subset of the usual product topology for XxS; not only are arbitrary non-linear continuous functions permitted, but even singularities outside compact sets, e.g. "at infinity." "Affine" tensor product gives $\delta:X\otimes_k S\to S$ in Aff_k iff of the form $a(x,s)+b(x)+c(s)+s_0$, with a bilinear, b,c linear, and $s_0\varepsilon S$. This series can be thought of as the following truncation of the Taylor series expansion of a non-linear function, $xAs+Bx+Cs+s_0$ with A,B,C matrices, so a vector; an improvement of the usual truncation by two (non-linear) terms.

Indications are (from Goguen[22]) the theory of such systems is quite as nice as that of the usual linear systems! For an affine machine with $X=K^n$ the direct analogue of X^* is $\coprod^{A}_{t\geq 0} \textcircled{A}^{t}_{K} K^n$, where an affine coproduct $\coprod^{A}_{t\geq 0} X_t$ is $(\coprod_{t\geq 0} A_t) \times (\coprod_{t>0} K)$ and $A \textcircled{A}_k B$ is $A \otimes_k B + A + B$ (with $\otimes_k$ the usual tensor product). Then $\textcircled{A}^{t}_{k} K^n$ has dimension $(n+1)^{t+1}-1$. Such results lead to an explicit formula for E(M). Note that every linear machine is affine, so this theory properly contains that of linear systems. But surprisingly, minimal affine realizations can be cheaper than minimal linear realizations! In analog computer terms, an affine system consists of scalors, summors, delayors, constants, and binary multipliers between inputs and states only.

The existence of minimal state objects for affine and Kelley systems, unique up to isomorphism, seems to be a very encouraging portent for a general theory of non-linear systems, although it is still some distance to what must be the main goal, algorithms for finitely constructing realizations.

We mention also the elegant "X-dynamics" approach of Arbib-Manes[23], "addressed machines" of Bainbridge[24], and pseudoclosed categories of Ehrig-Kreowski[19].

BIBLIOGRAPHY

1. Thatcher, J. W. "Characterizing Derivation Trees of Context-Free Grammars Through a Generatlization of Finite Automaton Theory," <u>J. Comp. and Sys. Sci. 1</u>, pp. 317-322, 1967.

2. Thatcher, J. W. and Wright, J. B. "Generalized Finite Automata With an Application to a Decision Problem of Second-Order Logic," <u>Math. Sys. Th. 2</u>, pp. 57-81, 1968.

3. Eilenberg, S. and Wright, J. B. "Automata in General Algebras," <u>Infom. Control 11</u>, pp. 52-70, 1967.

4. Goguen, J. A., Thatcher, J. W., Wagner E. G. and Wright, J. B. "A Junction Between Computer Science and Category Theory, I: Basic Concepts and Examples (Part 1)," Report RC 4526, IBM Watson Research Center, September 1972. (Part 2), to appear.

5. Mac Lane, S. and Birkhoff, G. <u>Algebra</u>, Macmillan, 1967.

6. Maibaum, T. S. E. "The Characterization of the Derivation Trees of Context-Free Sets of Terms as Regular Sets," <u>Proceedings 13th Annual Symposium Switching and Automata Theory</u>, College Park, Md., pp. 224-230, 1972.

7. Turner, R. "An Infinite Hierarchy of Term Languages - An Approach to Mathematical Complexity," Automata, Languages and Programming (Proc. Symp. IRIA, 1973) ed. M. Nivat, North Holland, pp. 593-608, 1973.

8. Wand, M. "An Algebraic Formulation of the Chomsky Hierarchy," Proc. of the First International Symposium: Category Theory Applied to Computation and Control, 1973.

9. Knuth, D. "Semantics of Context-Free Languages," Math Sys. Th. 2, pp. 127-145, 1968.

10. Naur, P. et al "Revised Report on the Algorithmic Language ALGOL 60," Communications of the ACM 6, pp. 1-17, 1963.

11. Scott, D. "The Lattice of Flow Diagrams," Symp. on Semantics of Algorithmic Languages, Lecture Notes in Math V. 188, Springer Verlag, 311-366, 1970.

12. Scott, D. and Strachey C. "Toward a Mathematical Semantics for Computer Languages," Proc. Symp. Computers and Automata, Polytechnic Inst. of Brooklyn, pp. 19-46, 1971.

13. Goguen, J. A. "On Homomorphisms, Simulations, Correctness and Subroutines for Programs and Program Schemes," Proc. 13th IEEE Symposium on Switching and Automata, College Park, Md., pp. 51-60, 1972.

14. Burstall, R. M. "An Algebraic Description of Programs with Assertions, Verification and Simulation," Proceedings of the ACM Conference on Proving Assertions About Programs, Las Cruces, N. M., pp. 7-14, 1972.

15. Floyd, R. W. "Assigning Meanings to Programs," Proc. Symp. Appl. Math 19 American Math. Soc., pp. 19-32, 1967.

16. Manna, Z. "The Correctness of Programs," J. Comp. and Sys. Sci. 3, pp. 119-127, 1969.

17. Burstall, R. M. and Thatcher, J. W. "The Algebraic Theory of Recursive Program Schemes," Proc. of the First International Symposium: Category Theory Applied to Computation and Control, 1973.

18. Goguen, J. A. "Minimal Realization Situations," unpublished manuscript from 1969; to appear in ADJ.

19. Ehrig, H. and Kreowski, J. J. "Power and Initial Automata in Pseudo-Closed Categories," Proc. of the First International Symposium: Category Theory Applied to Computation and Control, 1973.

20. Lawvere, F. W. "Functorial Semantics of Algebraic Theories," Proc. Nat. Acad. Sci. 50, pp. 870-872, 1963.

21. Goguen, J. A. "Discrete-Time Machines in Closed Monoidal Categories, I," Quarterly Report #30, Inst. for Computer Res., University of Chicago, Section IIB, 1971; To appear in Jnl. Comp. and Sys. Sci., Summarized in "Minimal Realization of Machines in Closed Categories," Bull. Amer. Math. Soc. 78, pp. 777-783, 1972.

22. Goguen, J. A. "Discrete-Time Machines in Closed Monoidal Categories II," in preparation.

23. Arbib, M. A. and Manes, E. G. "Machines in a Category: An Expository Intraduction," to appear in the SIAM Review 16, 1974.

24. Bainbridge, E. S. "Addressed Machines and Duality," Proc. of First International Symp.: Category Theory Applied to Computation and Control, 1973.

SCATTERING THEORY AND NON LINEAR SYSTEMS

F. Joanne Helton
Mathematics Department
Dowling College
Oakdale, New York 11769

J. William Helton
Mathematics Department
University of California
Los Angeles, California 90024

Systems theory is the study of an input/output situation and in a general sense scattering theory is also the study of an input/output situation. The evolution of systems theory and the evolution of scattering theory have been reasonably independent; however, the two are at their core quite closely related. What the authors have done is to show how to formulate systems theory in a scattering theory way. What we develop is one common setting in which to place abstract scattering theory and non linear systems theory.

In the standard Kalman theory one always associates with a system its frequence response function and its controllability and observability maps. In scattering theory one associates with a problem its scattering matrix and its forward and backward wave operators. If one looks at these objects in the formal setting developed here, then they can be compared. The frequency response function and the scattering operator are closely related as are the controllability and observability operators and the wave operator. This is described only briefly in this abstract the bulk of which is devoted to motivating and describing the formal structure. For a fuller account see [H-H].

It is reasonable to expect that cross fertilization between systems theory and scattering theory will be useful. Possibly one viewpoint will be more fruitful in particular circumstances than the other. For example, it is possible that in some distributed systems wave operators will be easier to compute and use. On the other hand systems theory might have a broadening influence on abstract scattering theory. Systems theory is based on a larger class of examples and could well serve as solid motivation for very general approaches to scattering theory.

The closest relative in scattering theory to state space systems theory is Lax-Phillips scattering theory and we give a simplified account of the ideas involved there [L - P_1] [L - P_2]. The vehicle for our discussion is a typical example. Given an obstacle which is translucent to sound, the problem is to determine the construction of the obstacle. This is done by sending sound waves toward the obstacle and then measuring the reflected waves at considerable distance from the obstacle. What the sound wave does near the obstacle is of no interest. Thus, each incoming sound wave can be identified with an outgoing sound wave. The map performing this identification is called the scattering operator for the particular obstacle. Thus, the basic problem is — given a 'scattering operator' find the obstacle. It is called the <u>inverse scattering problem</u>.

A reasonable way to formulate this situation is the following: Consider a wave at some fixed time t_0 and record its amplitude $u(x,t_0)$ and velocity $u_t(x,t_0)$ at a point x. There is a well defined map $U(t)$ of function pairs to function pairs so that

$$U(t)(u(x,t_0),u_t(x,t_0)) = (u(x,t_0+t),u_t(x,t_0+t)).$$

For example, this describes the behavior of a sound wave as time evolves.

Henceforth, assume that we have one particular obstacle and denote the time evolution of sound in the exterior of this obstacle by $U(t)$, it is called the <u>perturbed evolution</u>. If no obstacle exists we denote the time evolution by $U_0(t)$ and call it the <u>free evolution</u>. The space of amplitude-velocity pairs forms a Hilbert space denoted by H when given a suitable inner product. Henceforth, we shall work with discrete time. Assume each vector h in H can be regarded as a square summable infinite tuple $h = (\ldots,n_{-2},n_{-1},n_0,n_1,n_2,\ldots)$ with entries n_j in a Hilbert space N and with the property that $U_0(m)$ shifts h exactly m units to the right. This is true in many physical examples, for example acoustic or electromagnetic scattering in three dimensional space [L - P$_1$]. Now it can be easily shown that there exists (see [H-H]) an integer T and operators Z, B, C and D so that $U(1)$ when applied to a typical sequence h gives a sequence $(\ldots,n_{-2},n_{-1},Zq+Bn_{-1},Cq+Dn_{-1},n_{T+1},n_{T+2},\ldots)$ where $q = (\ldots,0,\ldots,0,n_0,\ldots,n_{T-1},0,\ldots,0,\ldots)$. Thus there is a system implicit in the original scattering problem

$$q' = Zq + Bn \qquad n' = Cq + Dn.$$

The state space Q of this system is the set of all such tuples q and the input and output space of this system is N. Although Z plays a prominent role in Lax-Phillips scattering theory the input and output operators B, C and D have never appeared in their work.

One can reverse this process and treat systems from a scattering viewpoint. We will study a simple but non linear system S with inputs and outputs in X and states in Q. S is defined by functions δ and λ which specify its behavior at time t as

$$q_{t+1} = \delta(q_t,i_t) \qquad o_t = \lambda(q_t,i_t)$$

where q_t is the state, i_t is the input and o_t is the output of A at time t. Frequently, in practice, there will be a distinguished letter $-$ in X with the property that $\delta(-,-) = -$ and $\lambda(-,-) = -$. In typical linear systems $-$ corresponds to 0.

The setting for what we call the scattering theory viewpoint is as follows. We define the universe Ω' of a system (this is not a scattering theory term) to be the set of all infinite tuples consisting of (input string, state, output string). Precisely Ω' is all tuples of the form

$$(1) \qquad (\ldots,i_{-3},i_{-2},i_{-1},\boxed{q},o_1,o_2,o_3,\ldots)$$

where i_{-j} and o_j are in X and q is in Q. The state will always be singled out by a box. Define Ω'_0 (respectively Ω'_+ and Ω'_-) to be all elements of Ω' with only finitely many entries (respectively positive entries and negative entries) not equal to -. It is natural to view the system S as acting on Ω'. We call the map $W : \Omega' \to \Omega'$ which describes how S acts on Ω' in each time interval the <u>evolution</u> map. It maps a typical sequence (1) to

$$(\ldots, i_{-3}, i_{-2}, \boxed{\delta(q, i_{-1})}, \lambda(q, i_{-1}), o_1, o_2, \ldots).$$

An important example is a free system. For a free system F the alphabets Q, X and Y are equal and F's action on a string is just to shift it to the right. Let Ω denote the infinite tuples for F. The evolution map for F acting on Ω is denoted by S_+. It maps (1) into

$$(\ldots, i_{-2}, \boxed{i_{-1}}, q, o_1, o_2, o_3, o_4, \ldots).$$

Note that the free machine is a reset machine. The inverse of S_+ is denoted by S_-; it maps (1) into

$$(\ldots, i_{-3}, i_{-2}, i_{-1}, q, \boxed{o_1}, o_2, \ldots).$$

The main theme is scattering theory is to compare an arbitrary system to a free system to get information about the arbitrary system. Since $X \subset Q$ one can think of the universe of the arbitrary system as containing a universe for a free system, namely Ω is the set of all strings in Ω' whose state entry is in X. This setup is much like that in the scattering theory described above.

The three basic objects associated with a classical scattering problem are the scattering operator Σ and the forward and backward wave operators W_+ and W_-, defined by $W_- = \lim_{t\to-\infty} U(-t)U_0(t)$, $W_+ = \lim_{t\to+\infty} U(-t)U_0(t)$ and $\Sigma = W_2W_-S_-$ where $W_2 = W_-^{-1}$. These constructs also make good sense for systems. We will compute the analog of W_- which is $\lim_{n\to\infty} W^nS_-^n$. The maps $W^nS_-^n$ are easily computed, for example $W^2S_-^2$ of a typical sequence (1) from Ω is

$$(\ldots, i_{-2}, i_{-1}, \boxed{\delta(\delta(o_2, o_1), q)}, \lambda(\delta(o_2, o_1), q), \lambda(o_2, o_1), o_3, o_4, \ldots).$$

If $W^n S_-^n$ is applied to a sequence of the form $w = (\ldots, i_{-1}, \boxed{q}, -, -, \ldots)$ in Ω_+ the resulting sequence is $(\ldots, i_{-1}, \boxed{\delta(-,x)}, \lambda(-,x), -, -, \ldots)$ regardless of n and we call this the limit of $W^n S_-^n w$. A limit exists in the same sense for any sequence in Ω_+ and is easily written down. Now $W_2 = \lim_{n\to\infty} S_-^n W^n$ can be computed similarly on Ω' and then W_2 along with W_- can be used to determine Σ on Ω_+ (see [H-H]).

Proposition 2, section 3 of [H-H] shows that the controllability map is just W_- projected onto the state space and that the observability map is just W_2 restricted to the state space. Also Σ has an analog in linear systems theory, namely the frequency response function. The realizability problem has a nice solution and scattering operators combine well under series and parallel composition so that the Krohn-Rhodes decomposition theorem has a simple formulation in terms of scattering operators.

We will now describe the scattering and wave operators for a very simple non linear system, an adding device. Let S be the system for which:

$$\delta(k,j) = - \qquad \delta(-,k) = k$$

$$\lambda(k,j) = k + j \qquad \lambda(-,k) = -$$

where $k, j = 1, 2, \ldots$.

1. The scattering operator Σ for this system applied to the sequence $a = (\ldots, a_{-1}, \boxed{a_0}, a_1, \ldots, a_n, -, \ldots)$ where $a_k \neq -$, $k < n$ gives the sequence $(\ldots, b_{-1}, \boxed{b_0}, b_1, \ldots, b_n, -, \ldots)$ where $b_n = -$, $b_{n-1} = a_n + a_{n-1}$, $b_{n-k} = -$ if k is even and $b_{n-k} = a_{n-k+1} + a_{n-k}$ if k is odd.

2. The backward wave operator W_- applied to the sequence a gives the sequence $(\ldots, a_{-1}, \boxed{c_0}, c_1, c_2, \ldots, c_{n+1}, -, \ldots)$ where $c_0 = -$ or b_0 according as n is odd or even and $c_k = b_{k-1}$, $k = 1, \ldots, n+1$.

3. The operator W_2 applied to the sequence a gives the sequence $(\ldots, d_{-2}, d_{-1}, \boxed{d_0}, a_1, a_2, \ldots)$ where $d_0 = a_0 + a_{-1}$, $d_{-k} = -$ if k is odd and $d_{-k} = a_{-k} + a_{-k-1}$ if k is even.

REFERENCES

[H-H] F. J. Helton and J. W. Helton, Scattering Theory for Computers and Other Non-linear Systems, submitted for publication.

[L-P_1] P. D. Lax and R. S. Phillips, "Scattering Theory", Academic Press, New York (1967).

[L-P_2] P. D. Lax and R. S. Phillips, Scattering Theory for Dissipative Hyperbolic Systems, Journal of Functional Analysis, Oct. 1973, Vol. 14, No. 2.

SYNTHESIS AND COMPLEXITY OF LOGICAL SYSTEMS

Hans-Jürgen Hoehnke
Akademie der Wissenschaften der DDR

Zentralinstitut für Mathematik und Mechanik
(DDR)108Berlin, Mohrenstr. 39

The description of structural synthesis and complexity of logical systems requires a formal language ("language of terms") in which logical systems can be recasted, as well as a suitable measure of complexity. Here only the first question is considered. With respect to the second question we are satisfied with the remark that the number of (eventually valuated) occurrences of building elements is often used as a measure of the complexity of a logical system (Shannon, Lupanov).

As logical systems we only treat here combinatorial and sequential networks of automata. Our investigations can be extended to derivations of grammars of formal languages (semithue systems) and to formal proofs, cf. J. Lambek (1973) and the author (1974).

We distinguish four cases of systems synthesis:

Number of basic alphabets	Combinatorial automata	Sequential automata
1	I	II
>1	III	IV

In the cases I and II, for practical reasons, the two-element alphabet {0,1} is usually chosen. In these cases only we have only one concept of synthesis. In case I an algebraic language of terms can easily be constructed. This holds also in case II for at least one language.

Now we consider case III in more detail and assume a system of building elements $\underline{F} = \langle f_g \rangle_{g \in G}$ of finitary functions

$$f = f_g : A_{i_1} \otimes \ldots \otimes A_{i_{n_g}} \longrightarrow A_{i_o} \quad (i_k \in J)$$

over a system $\underline{A}$ of basic alphabets A_i $(i \in J)$. (The A_i are sets, at first not assumed to be finite, and $A \otimes B$ denotes the cartesian product of sets A and B.) Then the pair $\Phi = \langle \underline{A}, \underline{F} \rangle$ is a heterogeneous algebra of type $\langle n_g \rangle_{g \in G}$ in the sense of Birkoff-Lipson (1970) with the carriers A_i and the fundamental operations f_g $(g \in G)$.

Let be $\underline{F}_{\underline{A}}$ the set of all functions

(1) $$f: A_{j_1} \otimes \ldots \otimes A_{j_n} \longrightarrow A_{j_o} \quad (j_k \in J)$$

over $\underline{A}$. On the set $\underline{F}_{\underline{A}}$ there are defined in the usual way "superpositions", i.e., certain full zero-, one- and two-placed algebraic operations (identity maps $1_j : A_j \longrightarrow A_j$, permutations of variables, identifications of variables, adjunction of fictive variables, and substitution of one function instead of the first variable of another function). The strait clone $O'(\Phi)$ of the algebra Φ then is defined as the closed hull (with respect to these superpositions) of the system $\underline{F}$ in $\underline{F}_{\underline{A}}$.

To this closure-process there corresponds a so-called language of operator terms which can be easily constructed. The variables of an operator term are interpreted not as elements of the sets A_i, but as elements of $\underline{F}_{\underline{A}}$, expecially of $\underline{F}$.

Let be $\underline{F}_{\underline{A}}$ decomposed according to

$$\underline{F}_{\underline{A}} = \bigcup_{j_o, j_1, \ldots, j_n \in J} \text{Ens}\,(A_{j_1} \otimes \ldots \otimes A_{j_n}, A_{j_o})$$

into subsets, where Ens $(A,B) = \{f: A \longrightarrow B\}$ consists of all maps of the form $f: A \longrightarrow B$. Then the superpositions appear as full operations over the sequence of sets

$$\text{Ens}\,(A_{j_1} \otimes \ldots \otimes A_{j_n}, A_{j_o}), \quad j_o, \ldots, j_n \in J, \ n \geq 0.$$

With these carrier-sets and with the operations of superposition the set $\underline{F}_{\underline{A}}$ as well as the subalgebras $O'(\Phi)$ of $\underline{F}_{\underline{A}}$ satisfy certain identites, and one may consider all algebras of the same type as $\underline{F}_{\underline{A}}$, satisfying all these identities. These algebras are called abstract strait clones. We can prove:

1) Every abstract strait clone is isomorphic to a "concrete" strait clone $O'(\Phi)$.

2) The class of abstract strait clones contains a free algebra. This algebra is a language of operator terms in case III.

In case I the operations defined above permit expression of the operation ∂ of omitting a fictive variable. For example, let $f(x_1,x_2) = f'(x_1)$; $x_1, x_2 \in A$, i.e.,

$$f = f' \otimes t: A \otimes A \longrightarrow A,\ f': A \longrightarrow A,\ t: A \longrightarrow A^o,\ A^o = \{\emptyset\}.$$

Then

$$f: = f' = d(f' \otimes t) = df, \quad \text{where} \quad d: x \longmapsto (x,x) : A \longrightarrow A \otimes A.$$

Multiplication of f by d corresponds to identification of variables x_1 and x_2 in $f(x_1,x_2)$.

On the other hand, let $f(x_1,x_2) = f'(x_1)$ be a function with $x_1 \in A$, $x_2 \in B$ and $A \neq B$. Then in general $f': A \longrightarrow A$ cannot be expressed in terms of f and superpositions.

Therefore in the cases of different alphabets A_i, the omission of fictive variables necessitates new partial operations

$$\partial = \partial_{j_1,\ldots,j_n,j_o} : \mathrm{Ens}\,(A_{j_1} \otimes \ldots \otimes A_{j_n}, A_{j_o}) \rightarrowtail \mathrm{Ens}\,(A_{j_2} \otimes \ldots \otimes A_{j_n}, A_{j_o})$$

(where $\rightarrowtail$ denotes a map "not everywhere defined on ... into"), the domain of definition of which contains all functions (1), which do not depend essentially from their first place, i.e., for which there is a function

$$f': A_{j_2} \otimes \ldots \otimes A_{j_n} \longrightarrow A_{j_o}$$

which makes the following diagram commutative.

$$\begin{array}{ccc} A_{j_1} \otimes A_{j_2} \otimes \ldots \otimes A_{j_n} & \xrightarrow{\quad f \quad} & A_{j_o} \\ \Big\downarrow p & & \Big\downarrow 1 \\ A_{j_2} \otimes \ldots \otimes A_{j_n} & \xrightarrow[\quad f' \quad]{} & A_{j_o} \end{array}$$

Here p denotes the canonical projection.

In this case we write $\partial f = f'$.

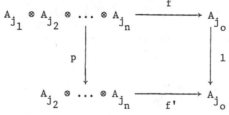

The decomposed set $\underline{F}_A$ together with the operations of superposition and the operations ∂ forms a special case of partial heterogeneous algebras in the sense of Kaphengst-Reichel (1972), simply called KR-algebras.

The KR-subalgebras of the KR-algebra $\underline{F}_A$ are considered by R. Pöschel (1973) under the name "Post *algebra". In the case of only one basic alphabet (card J = 1) the concepts "strait clone" and "Post *algebra" coincide (cf. above).

As the thoery of KR-algebras shows, there is a language of operator terms for Post *algebras, i.e. for those strait clones of $\underline{F}_A$, which are closed also with respect to the operations ∂.

In a similar manner one can establish a language of operator terms for the structural synthesis of sequential automata including feedback operation as a partial operation, cf. Budach-Hoehnke (1974).

REFERENCES

1. Birkhoff, G., and Lipson, J.D.: Heterogeneous algebras, J. Combinatorial Theory 8, 1970, 115-133.

2. Budach, L., and Hoehnke, H.-J.: Automaten und Funktoren, Berlin, 1974.

3. Higgins, P.J.: Algebras with a scheme of operators, Math. Nachr. 27, 1963, 115-132.

4. Hoehnke, H.-J.: Struktursätze der Algebra und Kompliziertheit logischer Schemata, Math. Nachr. (im Druck), 1974a.

5. Hoehnke, H.-J.: Kompliziertheit Boolescher Schaltkreise, Math. Nachr. (im Erscheinen), 1974b.

6. Kaphengst, H., and Reichel, H.: Operative Theorien und Kategorien von operativen Systemen, Studien zur Algebra und ihre Anwendungen, Berlin, 1972.

7. Kobrinski, N.E., and Trachtenbrot, B.A.: Einführung in die Theorie endlicher Automaten, Berlin, 1967.

8. Lambek,.J.: Functional completeness of cartesian categories, Preprint, Montréal, 1973.

9. Pöschel, R.: Postsche Algebren von Funktionen über einer Familie endlicher Mengen, Z. Math. Logik Grundlagen Math. 19, 1973, 37-74.

STRUKTURELLE VERWANDTSCHAFTEN VON SEMI-THUE-SYSTEMEN

Günter Hotz
Angewandte Mathematik
und Informatik

Universität des Saarlandes
66 Saarbrücken, W-Germany

Semi-Thue-Systems and Chomsky Grammars (C.L.) are associated with free monoidal categories (X-categories) in a natural way [Ho,Be]. We specialize the weak equivalence of C.L. by introducing different relationships $R(\tilde{F})$ based on the structure of the associated X-categories. These relationships are generated by chains of functors from a functor category $\tilde{F}$. The decision problem if two grammars are $R(\tilde{F})$-equivalent is strongly connected to the question if products in $\tilde{F}$ exist and if functors $\phi \in \tilde{F}$ are surjective. In this paper we report on some results which we have gained about this topics in the last years.

Semi-Thue-Systeme (STS) haben durch die Arbeiten von N. Chomsky für die Linguistik eine hervorragende Bedeutung erlangt. Der Ansatz zur Beschreibung der Syntax von Programmiersprachen in der Backus-Notation stellt vielleicht eine noch wichtigere Anwendung dieser Systeme dar. Durch Chomsky haben die (STS) eine kombinatorische Klassifikation erfahren, an die sich eine ausführliche Diskussion des Sprachumfangs dieser Sprachklassen anschloß. Ein großer Teil der Ergebnisse ist negativer Art: Es existieren zur Lösung der interessierenden Probleme keine universellen Algorithmen. Neben dem Wortproblem hat die Frage nach der Gleichheit zweier Chomsky-Sprachen (schwache Äquivalenz) eine besondere Aufmerksamkeit erfahren. Resultate über die strukturelle Verwandtschaft verschiedener Sprachen existieren kaum. Selbst bei der Herleitung von Normalformentheoremen für Grammatiken hat man sich mit der Feststellung der schwachen Äquivalenz der zugehörigen Sprachklassen begnügt. Strukturelle Aussagen enden im allgemeinen mit der Diskussion der Mehrdeutigkeitsfragen der Grammatiken.
Die Bedeutung dieser Fragestellung ist aber eminent wichtig. Findet doch die Interpretation der gesprochenen Sätze großteils, und die der Programme einer Programmiersprache fast ausschließlich, über die durch die Grammatiken definierte Ableitungsstruktur des Satzes, bzw. des Programms statt. Dies ist die Motivation der Untersuchungen, über die hier kurz berichtet werden soll.
Hierbei gehen wir auf Normalformtheoreme nicht ein, sondern verweisen in diesem Zusammenhang auf eine jüngst erschienene Arbeit [Ha] und das Buch [Ho-Cl], das die Herstellung von Normalformen unter kategoriellem Aspekt betrachtet.
Da es uns im folgenden vorwiegend auf die Herausarbeitung der Ideen ankommt, werden wir weder größte Allgemeinheit noch Vollständigkeit anstreben.
Sei $S = (A,P,\rightarrow)$ ein (STS), d.h. A eine endliche Menge, A^* freies Monoid über A und $P \subset A^* \times A^*$ ebenfalls endlich. Die Relation P wird in bekannter Weise zu der Relation $\rightarrow$ auf A^* fortgesetzt. Gilt für $w,v \in A^*$ die Beziehung $w \rightarrow v$, dann kann man nach allen Verfahren F fragen, $w \rightarrow v$ aus P zu konstruieren, Bezeichnung: $w \overset{F}{\rightarrow} v$. Jede solche Konstruktion F kann durch einen Ausdruck in o und x über P beschrieben werden, indem man definiert

$w \overset{F}{\to} v,\ v \overset{G}{\to} u$

$\Rightarrow w \overset{G \circ F}{\to} u$

und

$w \overset{F}{\to} w',\ v \overset{G}{\to} v'$

$\Rightarrow wv \overset{G \times F}{\to} w'v.$

Rechnet man mit diesen Ausdrücken unter Verwendung der Regel

$(F_1 \times F_2) \circ (G_1 \times G_2)$
$= (F_1 \circ G_1) \times (F_2 \circ G_2).$

Falls $F_1 \circ G_1$ und $F_2 \circ G_2$ definiert sind, dann bilden die Klassen, die man durch Zusammenfassung der Ausdrücke erhält, die durch Anwendung dieser Rechenregel ineinander überführt werden können, eine "freie" monoidale Kategorie [Mc 1], die wir als X-Kategorie [Ho 1] $\mathcal{F} = \mathcal{F}(\mathcal{S})$ bezeichnen. $\mathcal{S}$ und $\mathcal{F}$ bestimmen sich gegenseitig eindeutig. In $\mathcal{F}$ ist die "Struktur von $\mathcal{S}$" eingefangen. Unter strukturellen Aussagen über $\mathcal{S}$ verstehen wir Aussagen über $\mathcal{F}$.
Seien nun etwa $\mathcal{S}_1$ und $\mathcal{S}_2$ (STS) und $\mathcal{F}_1$ und $\mathcal{F}_2$ die zugehörigen X-Kategorien. $\mathcal{S}_1$ und $\mathcal{S}_2$ kann man als <u>elementar verwandt</u> bezeichnen, wenn gilt:
$\mathcal{F}_1$ ist eine Unter-X-Kategorie von $\mathcal{F}_2$: $(\mathcal{F}_1 \subset \mathcal{F}_2)$.

Eine weitere <u>elementare</u> Verwandtschaft kann in der Existenz von Funktoren
$\phi : \mathcal{F}_1 \to \mathcal{F}_2$
bestehen, das heißt in strukturerhaltenden Abbildungen von $\mathcal{F}_1$ in $\mathcal{F}_2$.
Diese Begriffe werden durch in der Kategorientheorie übliche Differenzierungen verfeinert. Wir formulieren unsere Resultate, ohne solche zusätlichen Forderungen im einzelnen anzugeben.
Verwendet man Semi-Thue-Systeme zur Definition formaler Sprachen
$L(\mathcal{S},T,s) = \{w \in T^* \mid s \to w\}$
mit
$T \subset A, s \in A - T, P\ (A - T)^* \times A^*$,
so wird man an die obigen Verwandtschaftsbeziehungen noch Forderungen stellen, die s und T^* betreffen.
Wir definieren nun ohne vollständig zu sein
$\mathcal{L} = (\mathcal{S},T,s)$ und $\mathcal{L}' = (\mathcal{S}',T',s')$
heißen <u>verwandt</u>, wenn es eine Kette $\mathcal{L} = \mathcal{L}_1, \mathcal{L}_2, \ldots, \mathcal{L}_k = \mathcal{L}'$ gibt, so daß $\mathcal{L}_i$ und $\mathcal{L}_{i+1}$ oder $\mathcal{L}_{i+1}$ und $\mathcal{L}_i$ für $i = 1,\ldots,k-1$ elementar verwandt sind im obigen Sinne.
Es stellt sich die Frage:

<u>Problem 1</u>

Unter welchen Voraussetzungen an $\mathcal{S}$ und die elementaren Verwandtschaften kann man die Verwandtschafts-Ketten verkürzen? Unter welchen Voraussetzungen sind stets Verkürzungen auf Längen $\leq m$ möglich, wo m unabhängig von den Paaren $\mathcal{L},\mathcal{L}'$ ist.

Problem 2

Man gehe das Problem 1 unter dem Gesichtspunkt der effektiven Konstruierbarkeit oder - noch schärfer - unter dem Gesichtspunkt der Komplexität an.
Es liegen zu beiden Problemen einige Resultate vor, die jetzt kurz geschildert werden.

Man schränkt die Fragestellung auf einseitig lineare (STS) ein, dann kann man zeigen:
Ist
$L(S,T,s) = L(S',T,s')$,
dann sind
$L = (S,T,s)$ und $L' = (S',T',s)$
verwandt. Es gilt darüber hinaus:
Es gibt unter dieser Voraussetzung stets eine Kette
$L = L_1,\ldots,L_9 = L'$, so daß
$L_i = L_{i+1}$ und
L_i elementar verwandt zu
$L_{i\pm 1}$ ist [Ho 2].

Es liegen Ergebnisse vor, wann gewisse elementare Verwandtschaften effektiv entscheidbar sind [Ber], [Schn], [Ho 3].
Als Beispiel werde ein Satz aus [Ber] angegeben, der sich auf die Klasse der kontextfreien Systeme bezieht.
Für die Menge der Funktoren $\phi : F_1 \to F_2$, die durch Fortsetzung einer Abbildung $\phi' : P_1 \to F_2$ gewonnen werden, ist es entscheidbar, ob ϕ surjektiv oder injektiv ist.

Dieser Satz stellt eine Verallgemeinerung eines Resultats aus [Schn] dar, wo der Satz nur für nicht "erweiternde" Funktoren bewiesen wurde und eines Satzes aus [Ho 3], wo dieser Satz unter Einschränkung auf lineare Systeme gezeigt wurde. Für kontextsensitive Systeme sind diese Eigenschaften nicht mehr generell entscheidbar.
Klassen von Elementarverwandtschaften, für die ähnlich scharfe Sätze wie der aus [Ho 2] zitierte Satz beweisbar sind, werden in [Ba-Ho] angegeben.

Die "Verkürzungssätze" sind Sätze über die Existenz von Produkten in Funktor-Kategorien der X-Kategorien. Diese Sätze stehen in engstem Zusammenhang mit der Frage nach der Zerlegbarkeit von Systemen S. Hierzu wurden Kriterien in [Cl-Wa] und [Schn-Wa] sowie [Wa] gewonnen. Ohne die Bedingung des Nichterweiterns an die Funktoren führen Produktkonstruktionen aus der Kategorie heraus. Man erhält Sprachen L mit card $P = \infty$.

REFERENCES

[Ba,Ho] Bartholomes F. und Hotz G.: "Homomorphismen und Reduktionen linearer Sprachen", Lecture Notes in Operations Research and Mathematical Systems, Vol. 32, 143 S., (1970).

[Be] Benson D.: "Syntax and Semantics: A Categorical View", Inf. and Control 17, 145-160, (1970).

[Ber] Bertsch E.: "Surjectivity of Functors on Grammars", Berichte des Sonderforschungsbereichs Elektronische Sprachforschung.

[Cl-Wa] Claus V. und Walter H.: "Zerlegungen von Semi-Thue-Systemen", Computing 4, 107-124, (1969).

[Gr] Greibach S. A.: "A New Normal Form Theorem For Contextfree Phrase Structure Grammars", Journal ACM 12, 45-52, (1965).

[Ha] Harrison M. A.: "On Covers and Precedence Analysis", GI-Jahrestagung Hamburg, 1973; Ed. G. Goos und J. Hartmanis: Lecture Notes in Computer Science Vol. 1, 1-17, Springer-Verlag, Berlin, Heidelberg, New York.

[Ho-Cl] Hotz G. und Claus V.: "Automatentheorie und Formale Sprachen III", Bibliographisches Institut Mannheim, 240 S., (1971).

[Ho 1] Hotz G.: "Algebraisierung des Syntheseproblems von Schaltkreisen" I und II; Elektronische Informationsverarbeitung und Kybernetik (EIK) 1, 185-231, (1965).
Hotz G.: "Eindeutigkeit und Mehrdeutigkeit formaler Sprachen", (EIK) 2, 235-246, (1966).

[Ho 2] Hotz G.: "Übertragung automatentheoretischer Sätze auf Chomsky-Sprachen", Computing 4, 30-42, (1969).

[Ho 3] Hotz G.: "Reduktionssätze über eine Klasse formaler Sprachen mit endlich vielen Zuständen", Math. Zeitschrift 104, 205-221, (1968).

[Ho 4] Hotz G.: "Erzeugung formaler Sprachen durch gekoppelte Ersetzungen", 4. Colloquium über Automatentheorie, München 1967. Ed. F. L. Bauer und K. Samelson, verlegt durch TU München.

[Mcl] MacLane S.: "Kategorien - Begriffssprache und mathem. Theorie", Springer-Verlag, Berlin, Heidelberg, New York, 295 S., (1972).

[Schn] Schnorr C.-P.: "Transformational Classes of Grammars", Inf. and Control 14, 252-277, (1969).

[Schn-Wa] Schnorr C.-P. und Walter H.: "Pullbackkonstruktionen bei Semi-Thue-Systemen", (EIK) 5, 27-36, (1969).

[Vo 1] Vollmerhaus W.: "Die Zerlegung von kontextfreien Semi-Thue-Systemen mit Anwendung auf das Analyseproblem kontextfreier Sprachen", Beiträge zur Linguistik und Informationsverarbeitung, 12, (1967).

[Vo 2] Vollmerhaus W.: "Über die Zerlegung von freien X-Kategorien", 4. Colloquium über Automatentheorie, München 1967, Ed. F. L. Bauer und K. Samelson, verlegt durch TU München.

[Wa 1] Walter H.: "Pullbackkonstruktionen bei Semi-Thue-Systemen", 4. Colloquium über Automatentheorie, München 1967, verlegt durch TU München.

[Wa 2] Walter H.: "Verallgemeinerte Pullbackkonstruktionen bei Semi-Thue-Systemen und Grammatiken", (EIK) 6, 239-254, (1970).

[Wa 3] Walter H.: "Einige Topologische Aspekte der syntaktischen Analyse und Übersetzung bei Chomsky-Grammatiken", erscheint in J. Comp. and System Science.

CONTROL OF LINEAR CONTINUOUS-TIME SYSTEMS DEFINED OVER RINGS OF DISTRIBUTIONS

Edward W. Kamen
School of Electrical Engineering

Georgia Institute of Technology
Atlanta, Georgia 30332, U.S.A.

1. Introduction

Since Kalman [1] presented the k[z]-module description of linear constant discrete-time systems, further research along this line has proceeded in two main directions: replacing the field of scalars k by rings and replacing k[z] by other rings of operators. In this paper we are concerned with only the latter direction which has been applied to the study of linear constant continuous-time systems, as developed by Kalman and Hautus [2] and Kamen [3,4,5]. Also, Wyman [6] has considered a categorical approach to the concept of systems over operator rings.

The operator theory of linear continuous-time systems consists of two different approaches. In the first, the F,G,H matrices of the state space description are defined over a ring of operators generated from a finite number of convolution operators corresponding to the components of the class of systems under consideration. This theory applies to classes of operational-differential systems referred to as hereditary systems (in which the derivative of the "state" at time t_o depends on the state and input over a past interval $(t_o-\tau,t_o]$ with τ fixed and $0 < \tau \leq \infty$). For example, when F,G,H are defined over an operator ring generated from a finite number of Dirac distributions (acting as delay operators), we obtain an algebraic framework for a class of delay-differential systems. The operator rings are always Noetherian; that is, every ideal is finitely generated. As a result, this algebraic approach yields a new type of finiteness in the mathematical representation of systems that are infinite-dimensional in the classical sense. For the details, see Kamen [7].

The second approach in the operator theory of continuous-time systems [2,3,4,5] is motivated by the following observations: in the usual mathematical representation of linear continuous-time systems, the input-output function spaces and the state space are linear spaces over the field of real numbers R. The basic idea in the operator theory is to extend R to a convolution algebra T of functions (acting as convolution operators), such that the input-output function spaces and the state space are T-modules and various system maps (such as the input-output map) are T-module homomorphisms. Extensions are possible when the system is constant because of the convolution expression relating inputs and outputs. The primary reason for considering this construction is that the state space can be finitely generated as a T-module when it is infinite dimensional as a R-linear space. The finiteness as a T-module yields new techniques for studying infinite-dimensional systems. For example, as demonstrated in this paper, we can apply the theory of finitely-generated modules to the problem of control.

One further comment is in order before we begin with this second operator approach: since the idea is to extend R, we want to work with a convolution algebra of functions T in which R can be embedded. For a natural setup, this means that T should contain δ_o = Dirac distribution at $\{0\}$ so that we have the embedding $R \to T: a \mapsto a\delta_o$. In other words, T should be a convolution algebra of distributions containing the multiplicative identity δ_o. Of course, the presence of δ_o is also necessary in order to consider various algebraic operations such as inversion.

2. System Definition

Let T be a convolution R-algebra of (Schwartz) distributions on R with compact support $\subset (-\infty,o]$, and $T \ni \delta_a$ = Dirac distribution at $\{a\}$, $a \leq o$.

Let $\Omega \supset T$ (resp. Γ) be a topological T-module of distribution on R with compact support $\subset (-\infty,o]$ (resp. distributions on (o,∞)) with $\Omega^m(\Gamma^k)$=m-fold (k-fold) direct sum with the usual structure. Finally, for any $\omega = (\omega_1,\ldots,\omega_m)^{TR} \in \Omega^m$, TR = transpose, define

$$\ell(\omega) = \min_{i=1,2,\ldots,m}\{a_i \in R: a_i = \min_t\{t \in \text{supp } \omega_i\}\} \leq 0.$$

We then have

<u>Main Definition</u>: A generalized linear constant continuous-time system over T is a sextuple $\Sigma = (\Omega^m, X, \Gamma^k, g, h, \Psi)$, where X is a topological T-module, $g:\Omega^m \to X$ and $h:X \to \Gamma^k$ are (topological) T-module homomorphisms, and Ψ is a map given by

$$\Psi:\Omega^m \times X \times R^- \to X: (\omega, x, a) \mapsto \delta_{a+\ell(\omega)} \cdot x + g(\omega)$$

where $R^- = \{a \in R: a < 0\}$ and $\cdot$ denotes multiplication in the T-module structure on X.

In the above definition, Ω^m, X, and Γ^k are called the input, state, and output modules respectively. Given $\omega \in \Omega^m$, $g(\omega)$ is the state of the system Σ at time t = 0 resulting from input ω; given $x \in X$, h(x) is the output response on (o,∞) resulting from state x at t = 0; and $\Psi(\omega,x,a)$ is the state of Σ at t = 0 due to input ω and initial state x at time $t = a + \ell(\omega)$. (The parameter a is necessary in order to consider those cases in which ω contains Dirac distributions at $\{\ell(\omega)\}$.) The composition $hg:\Omega^m \to \Gamma^k$ is called the input/output map of the system Σ.

As a consequence of the topological module structure, in most instances (see [4] for details) we can represent the map $f \triangleq hg$ <u>uniquely</u> by an element of the space $U^{k\times m}$ of $k\times m$ matrices over U = all distributions on R with support $\subseteq [o,\infty)$. That is,

$$f:\Omega^m \to \Gamma^k: \omega \mapsto (\Theta * \omega)|_{(o,\infty)}, \quad \Theta \in U^{k\times m}$$

where * = convolution and $|_{(o,\infty)}$ denotes restriction of components to (o,∞) in the sense of distributions.

It is pointed out here that even though the systems defined above are not specified by differential equations, it is possible to obtain deep results on system behavior by using the module structure. This is demonstrated in the next section wherein results on control are given.

3. Control

An important property of a system Σ defined over T is that the state module X may be finitely generated when X is infinite dimensional as an R-linear space. For example, if $T = \Omega$, Ω^m is a free T-module, and if g is onto (complete reachability), then X is finitely generated, and cyclic if $m = 1$. For simplicity, here we consider only the case when X is cyclic and $m = k = 1$ (see [5] for generalizations). Under this constraint, X is isomorphic to the quotient module T/A_X where A_X = annihilating ideal of X. We can compute A_X from the input/output map $f:\omega \mapsto (\theta*\omega)|_{(o,\infty)}$, $\theta \in U$, as follows. Let $\mathcal{E}'_{(-\infty,o]}$ denote the space of all distributions on R with compact support contained in $(-\infty,o]$. Then if $X = \Omega/\text{Ker } f$ and g = canonical map, we have

(1) $$A_X = \{\pi \in T:\theta*\pi \in \mathcal{E}'_{(-\infty,o]}\}.$$

For a proof of this and the following results, see [5].

We now make the following

Definition: Given a system Σ with $m = k = 1$, a state $x \in X$, $x \neq 0$, is controllable (to zero) relative to T if and only if there exists a $\pi \in T$ such that

$$\delta_\tau\cdot x + g(\pi) = 0, \quad \tau < \ell(\pi) = \min_t\{t \in \text{supp } \pi\}$$

We immediately have

Proposition 1: If $X \approx T/A_X$ with generator $g(\delta_o) \neq 0$ and $A_X = \{0\}$, then no nonzero state of X is controllable rel T.

When $A_X \neq \{0\}$, the controllability of states is directly related to the following division operation in T: Given $\alpha,\beta \in T$, α is said to divide β with quotient $\lambda \in T$ and remainder $\pi \in T$ if $\beta = \alpha*\lambda + \pi$.

Theorem 1: If $X \approx T/A_X$ with gen $g(\delta_o) \neq 0$ and $A_X = \alpha T$, then $x = \sigma \cdot g(\delta_o)$, $\sigma \in T$, is controllable (rel T) if and only if α divides $\delta_\tau * \sigma$ with a remainder $\pi \in T$ such that supp $\pi \subset (\tau, o]$, in which case $-\pi$ is a control signal.

The next result that we give does not require A_X to be principal. First, fix $\tau \in R$, $\tau < 0$, and let $d = \delta_\tau$. Define

$$T[d] = \{\sum_{i=o}^{n} \alpha_i * d^i : \alpha_i \in T\}, \; d^i = \delta_{i\tau}$$

Given $\chi = \sum_{i=o}^{n} \alpha_i * d^i \in T[d]$, let $\ell(c_\chi) = \min_{i=o,1,\ldots,n}\{\ell(\alpha_i)\}$. Then we have

Theorem 2: Given $X \approx T/A_X$ with gen $g(\delta_o) \neq 0$, if there exists a $0 \neq \chi = \sum_{i=o}^{n} \alpha_i * d^i \in T[d] \cap A_X, \alpha_n = \delta_o$, then $x = \sigma \cdot g(\delta_o)$ is controllable rel T in $|n\tau|$ seconds if $\tau < \ell(c_\chi) + \ell(\sigma)$, in which case the control is the solution $\pi \in T[d]$ of $\sigma * d^n + \pi = 0 \pmod{\chi}$ with supp $\pi \subset (n\tau, o]$.

Theorem 2 can be utilized to compute control signals. To illustrate this, we give the following

Example: Consider a delay-differential system with $T = \Omega = \mathcal{E}'_{(-\infty,o]}$, $X = T/\mathrm{Ker}\, f$, g = canonical map, and f generated from the input/output equation

$$3\,\frac{dy(t-1)}{dt} + y(t) = 5\omega(t-2), \; y = \text{output}, \; \omega = \text{input}. \tag{2}$$

Let $p = \delta_o^{(1)}$ = first derivative of δ_o and let $d = \delta_{-1}$, then (2) is equivalent to $(3d*p + d^2)*y = 5\omega$. By (1), we have that $(3d*p + d^2)T \subset A_X$.

Now suppose we want to drive the state $x = \sigma \cdot g(\delta_o)$ to zero where

$$\sigma = \begin{cases} t + .5, & -.5 < t < -.25 \\ -t, & -.25 < t < 0 \\ 0, & \text{otherwise} \end{cases}$$

If we take $\chi = 3d*p + d^2 \in T[d] \cap A_X$, by Theorem 2 we can drive x (at t=-2) to 0 in 2 seconds with control π given by $\sigma * d^2 + \pi = 0 \pmod{\chi}$.

By actually dividing χ into $\sigma * d^2$ in T[d], we obtain the control

$$\pi = 3dp\sigma = \begin{cases} 3, & -1.5 < t < -1.25 \\ -3, & -1.25 < t < -1.0 \\ 0, & \text{otherwise} \end{cases}$$

4. Discussion

The results given above show that the control of infinite-dimensional continuous-time systems can be studied in terms of an algebraic framework. The power of this approach is mainly a result of the finiteness of the state space as a T-module. Recent work [5] has shown that an even "richer" structure can be obtained by working with subrings S of an initial operator ring T with the rings S possessing various finiteness characteristics. The basic idea is that given a system Σ over T, for any subring $S \subset T$, we can view Σ as a system over S (i.e., we have a change of rings). For example, taking $S = R[\delta_0^{(1)}]$, which is a principal ideal domain, we obtain an algebraic theory for finite-dimensional systems. More generally, the case when T is a Noetherian domain leads to an algebraic theory of operational-differential systems [5]. Finally, it now appears that time-varying and nonlinear systems can also be studied in a similar algebraic framework by considering non-commutative operator rings (Kamen [8]).

REFERENCES

1. R. E. Kalman, "Algebraic Structure of Linear Dynamical Systems. I. The Module of Σ," Proc. Natl. Acad. Sci. (USA), Vol. 54, pp. 1503-1508, 1965.

2. R. E. Kalman and M. Hautus, "Realization of Continuous-Time Linear Dynamical Systems," in Proc. Conf. on Ordinary Diff. Equ., (Ed. L. Weiss), NRL Math. Research Center, pp. 151-164, June, 1971.

3. E. W. Kamen, "A Distributional-Module Theoretic Representation of Linear Continuous-Time Systems," Ph.D. Thesis, Stanford Univ., May, 1971.

4. ___________, "Representation of Linear Continuous-Time Systems by Spaces of Distributions," IRIA Report INF 7213/72016, France, June, 1972.

5. ___________, "Topological Module Structure of Linear Continuous-Time Systems," submitted to SIAM J. Control.

6. B. F. Wyman, "Linear Systems over Rings of Operators," this volume.

7. E. W. Kamen, "On an Algebraic Theory of Systems Defined by Convolution Operators," to appear in Math. Systems Thy.

8. ___________, "Algebraic Results on Time-Varying Systems," in preparation.

CELLULAR AUTOMATA WITH ADDITIVE LOCAL TRANSITION

Wolfgang Merzenich
Informatik

Universität Dortmund
46 Dortmund-Hombruch, Postfach 500, Germany

Since J. v. Neuman (1) introduced the model of cellular automata many papers published deal with several aspects of a formal theory of cellular automata (2, 3, 4, 5, 6, 7). A cellular automaton induces a function from the configuration space into itself. This function is called the global transformation, and it is a problem to decide for a given function whether or not it is a global transformation for some cellular automaton (8). In this paper we treat the special case for cellular automata with additive transition functions. In this case we can give a complete characterization of the additive endomorphisms of the configuration space, which are induced by additive cellular automata. These endomorphisms form a semiring that is isomorphic to the group-semiring R(D,End(S)). A result of Amoroso, Cooper (9) and Ostrand(10) can be shown to be a simple consequence of this characterization. We try to give an algebraic treatment using the language and some simple facts from category theory (12).

1 Basic definitions

First we give some general definitions from the theory of cellular automata (4)

Definition 1: Let D be an abelian group and $k \in \mathbb{N}_o$ (the set of nonnegative integers). A neighborhood index $N \in D^k$ (NB-index) on D is a k-tuple of distinct elements of D. The ordered pair (D,N) is called an <u>array</u> (or tessalation) of degree k

<u>Definition 2:</u> Let $S \neq \emptyset$ be a set, $k \in \mathbb{N}_o$ and $\tau : S^k \to S$ a k-ary operation on S. Then the algebra (S,τ) of type <k> is called a <u>local transition</u> and τ is called <u>transition function</u>

<u>Definition 3:</u> A cellular automaton (CA) is a 4-tuple (D, N, S,τ), where (D, N) is an array, (S,τ) a local transition and the degree of (D,N) is the same as the arity of τ.

An intuitive interpretation of this structure is the following: D is an index-set for a family $(G_x | x \in D)$ of automata and each G_x is a copy of the automaton G. The inputs of G are elements of S^k and S is the state-set of G. If $N = (a_1,\dots,a_k)$ automaton G_x is connected with the k neighbors $G_{x+a_1},\dots,G_{x+a_k}$ and takes the k-tuple of their states as input. If at time t (discrete time) G_x is in state s(x, t) then its next state is $s(x, t + 1) := \tau(s(x + a_i, t) \mid i = 1,\dots,k)$.

<u>Definition 4:</u> Let (D, N, S,τ) be a CA

i) A function $c : D \to S$ is called a <u>configuration.</u>

ii) Let A(D, S) be the set of all configurations with domain D and range S. A(D, S) is called <u>configuration space.</u>

A configuration can be thought of as the global state of an array of automata. In the category of sets A(D, S) is the product of S with index-set D (i.e. $A(D, S) = \underset{x \in D}{X} S = S^D$). So we have projection functions $P_x : A(D, S) \to S$ by $p_x\, c = c(x)$ for all $x \in D$. For $k \in \mathbb{N}_o$ let F_k be the functor from the category of sets into itself that maps a set A to the product A^k and a function $f : A \to B$ to $F_k(f) : A^k \to B^k$ by $p_i F_k(f) := f p_i$, i.e. componentwise f. If (D, N) is an array with $N = (a_1,\dots,a_k)$ we define the function $\phi : D \to D^k$ by $p_i\, \phi(x) = x + a_i$, i.e. ϕ assigns to each $x \in D$ its k-tuple of neighbors.

Definition 5: Let $M = (D, N, S, \tau)$ be a CA of degree k, then we define the map $H : A(D, S) \to A(D, S)$ by

$$H(c) = \tau F_k(c)\ \phi$$

H is called the global transformation of M.

We abbreviate A(D,S) simply by A and show that the global transformation H factorizes as follows

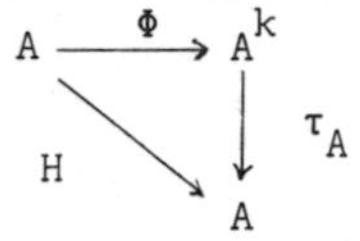

Where Φ is a k-tuple of translations of configurations and τ_A is the k-ary operation in the product S^D corresponding to the operation τ in S, i.e. $p_x \tau_A(c_i \mid i = 1,\dots,k) = \tau\ (p_x c_i \mid i = 1,\dots,k)$. The diagram expresses the following fact. To achieve the successor configuration of configuration c first produce k translations of c, according to the components of the NB-index N. Then apply the operation τ_A to this k-tuple of configurations.

2. Local transition with additional structure

The local transition was defined as an algebra $\underline{S} = (S, \tau)$ of type <k>. If we take a general algebra $\underline{S} = (S, \tau, (f_\mu \mid \mu \in I))$ with a k-ary operation τ and a family $(f_\mu \mid \mu \in I)$ of operations of type $<\gamma>$, i.e. f_μ is a $\gamma(\mu)$-ary operation. Let $\mathcal{K}$ be a class of algebras of the above type ($<k,\gamma>$) so that the direct procuct of members of $\mathcal{K}$ is again in $\mathcal{K}$. (e.g. let $\mathcal{K}$ be an equationally defined class (11)). By definition $\mathcal{K}$ is a category with homomorphisms of algebras as morphisms and $\mathcal{K}$ has products. If $(S, \tau, (f_\mu \mid \mu \in I)) = \underline{S} \in \mathcal{K}$ then $\underline{A} = \underline{S}^D = \underset{x \in D}{\mathsf{X}}\ \underline{S} \in \mathcal{K}$. We want to know wether the function $H : A \to A$ is a morphism in $\mathcal{K}$. Let $\mathcal{K}$ be defined as above.

Theorem 1: The global transormation $H : A \to A$ of a CA $(D, N, S, \tau,)$ is an endomorphism in $\mathcal{K}$ iff the following condition holds for τ

$$\tau(f_\mu(s_{ij}|i)|j) = f_\mu(\tau(s_{ij}|j)|i)$$

for all $\mu \varepsilon I$ and all arguments $((s_{ij}|i)|j)$ where $i \leq \gamma(\mu)$ and $j \leq k$

Explanation: $((s_{ij}|i)|j)$ is a k-tuple of $\gamma(\mu)$-tuples over S and $((s_{ij}|j)|i)$ is a $\gamma(\mu)$-tuples of k-tuple over S.

Because of this condition we now consider the case, where the additional structure on S is that of a commutative monoid, i.e. $\underline{S} = (S, \tau, +, o)$ where $(S, +, o)$ is a commutative monoid. Let $\mathcal{K}$ be the category of commutative monoids with the obvious morphisms. If S and T are objects of $\mathcal{K}$, we denote $[S, T] := \text{Mor}(S, T)$. If $\alpha, \beta \varepsilon [S, T]$ we define a sum of morphisms : $(\alpha + \beta)s = \alpha s + \beta s$ and a zero-morphism : $0s = o$. With these definitions $\mathcal{K}$ is a semiadditive category (12). We call the morphisms of $\mathcal{K}$ additive functions. Let End(S) denote $[S, S]$, the set of endomorphisms of S. End(S) is a semiring i.e. the additive structure is a commutative monoid; multiplication is the composition of functions. In $\mathcal{K}$ there exist products, coproducts and a zero object. We have the following relation between projections p_j and injections u_i

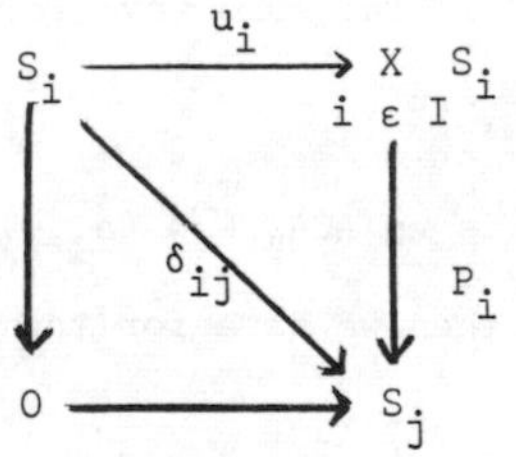

where δ_{ij} is defined as $\delta_{ij} := \begin{cases} 1_{s_i} & \text{if } i = j \\ o & \text{else} \end{cases}$

If $c \varepsilon \bigoplus_{i \varepsilon I} S_i$ is an element of the coproduct we have the representation : $c = \sum_{i \varepsilon I} u_i p_i c.$

For a CA (D, N, S, τ) and a commutative monoid $\underline{S}$ = (S, +, o) with the same set S we write $\underline{S}_{\tau}$: = (S, τ, +, o). The condition of Theorem 1 is equivalent to : τ is a morphism $\tau : \underline{S}^k \to \underline{S}$. Hence the global transformation H for the CA above is an endomorphism $H : \underline{S}^D \to \underline{S}^D$ in $\mathcal{K}$ iff τ is a morphism in $\mathcal{K}$, i.e. an additive function. In this case ($\tau \in [S^k, S]$) we say that the CA is an additive CA (ACA) and denote it by (D, N, $\underline{S}_{\tau}$). For an ACA τ has a matrix representation:

$$\tau(s_j \mid j) = \tau\left(\sum_i u_i p_i (s_j \mid j)\right) = \sum_i \tau u_i (s_i)$$

where $r_i := \tau u_i \in \mathrm{End}(S)$ and the matrix $(r_j \mid j)$ defines the function $\tau = \sum_i r_i p_i$.

To get a representation of configurations that depends on the group structure of the array we define for each $x \in D$ a translation function t_x for configurations by $p_y t_x := p_{y-x}$. It can be seen that the injections $u_x : S \to A$ can be factorized as follows: $u_x = t_x u_o$.
If $c \in \bigoplus_{x \in D} S$ then we can represent c as a sum:

$$c = \sum_{x \in D} u_x p_x c = \sum_{x \in D} t_x u_o c(x)$$

A definition of a sumable family of configurations allows us to represent the elements $c \in A$ ($= S^D$) in the same way. For $r \in \mathrm{End}(S)$ we define $\bar{r} \in \mathrm{End}(A)$ by: $p_x \bar{r} = r p_x$.

Theorem 2: If (D, N, $\underline{S}_{\tau}$) is an ACA, N = $(a_1,\dots,a_k)$ and τ represented by $(r_1,\dots,r_k)$ ($r_i \in \mathrm{End}(S)$), then we have for the global transformation H

$$H = \Sigma\, \bar{r}_j\, t_{-a_j}$$

The global transformation H is a sum of endomorphisms that are easily derived from the NB-index N and a representation of τ .

3 The class of global transformations

With these results we get an algebraic characterization of the class of all global transformations of a given configuration space A(D, S) that are induced by ACAs. Let this class be denoted by $\mathcal{A}(D, S)$. To this end we define the concept of DS-matrices.

Definition 6 : Let D be an abelian group and S an object of a semi-additive category with zero object. Let $m : D \times D \to \mathrm{End}(S)$ be a function into the semiring of endomorphisms with the property:

$$\bigwedge_{x \in D} \{y \in D \mid m(x, y) \neq o\} \text{ finite;}$$

then the family $(m(x,y)) : (m(x,y) \mid (x,y) \in D \times D)$ is called a DS-matrix.

The product $(p(x, y))$ of $(m(x, y))$ and $(n(x, y))$ is defined as

$$p(x, y) = \sum_{z \in D} m(x, z)\, n(z, y)$$ and is again a DS-matrix, and

$I : (\delta_{x,y})$ is the neutral element for multiplication. With componentwise addition the DS-matrices form a semiring $\mathcal{M}(D, S)$. A DS-matrix induces the following endomorphism $\bar{M} : S^D \to S^D$:

$$p_x \bar{M} = \sum_{y \in D} m(x, y)\, p_y$$

Global transformations of ACAs turn out to be special DS-matrices.

Theorem 3: A DS-matrix $(h(x, y))$ yields an endomorphism that is a global transformation of an ACA with configuration space $A(D, S)$ iff:

$$\bigwedge_{x,y \in D} h(x, y) = h(o, y - x)$$

This property is respected by the product operation and addition of DS-matrices. So the DS-matrices that represent global transformations of ACAs form a subsemiring of $\mathcal{M}(D, S)$. The elements of this semiring are completely determined by their o-rows (i.e. $(h(o, x) \mid x \in D)$) and so they can be represented by functions $h : D \to \mathrm{End}(S)$ with finite support. This semiring is well known as the group-semiring $R(D, \mathrm{End}(S))$ (13).

Theorem 4: The global transformations of ACAs on a given configuration space $A(D, S)$ form a semiring $\mathcal{A}(D, S)$ that is isomorphic to the group-semiring $R(D, \mathrm{End}(S))$.

As an example for application of this result we can give a simple proof of a theorem of Amoroso and Cooper (9) and Ostrand (10) in the theory of CA that is concerned with the self reproduction of configurations.

REFERENCES

1)v.Neumann, Theory of Self-Reproducing Automata, Urbana 1966

2) Burks, (ed.), Essays on Cellular Automata, Urbana 1970

3) Codd, Cellular Automata, New York 1968

4) Amoroso-Lieblein-Yamada, A Unifying Framework for the Theory of Iterative Arrays of Machines, ACM Symp., Theory of Comp., Conference Record Marina del Rey, Calif. May 5-7, 1969

5) Amoroso-Yamada, Tessalation Automata, Inf. and Contr. 14, 1969

6) Moore, Machine Models of Self-Reproduction, Proc. of Symp. in Appl. Math. vol. 14, AMS 1962

7) Smith III, Cellular Automata Theory, Stanford Univers., Techn. Rep.NO. 2, Dec. 1969

8) Richardson, Tessalations with Local Transformations, J. of Comp. and System Sciences 5, 1972

9) Amoroso-Cooper, Tessalation Structures for Reproduction of Arbitrary Patterns, J. of Comp. and System Sciences 5, 1971

10) Ostrand, Pattern Reproduction in Tessalation Automata of Arbitrary Dimension, J. of Comp. and System Sciences 5, 1971

11) Grätzer, Universal Algebra, Van Nostrand, Princeton N. J. 1968

12) Mitchell, Theory of Categories, New York - London 1965

13) Lambek, Lectures on Rings and Modules, Blaisdell Publ. Comp. Waltham Mass. 1966

AUTOMATA IN SEMIMODULE CATEGORIES

José Meseguer *
Electronics and Geometry

Ignacio Sols
Geometry

Zaragoza University
Faculty of Science, Zaragoza, Spain

In the present abstract we propose the framework of semimodule categories and their subcategories as a suitable common framework for various situations in Automata and Control theories, including in particular deterministic, stochastic, non-deterministic, partial, relational, bilinear and fuzzy sequential machines and switching networks. New relationships between all these classes of systems appear naturally in this common framework. In particular a common tensor calculus is introduced. The presentation of switching networks is based in Pfender's preprint (1). The authors are indebted to Professors R. Moreno-Díaz and J.L. Viviente of Zaragoza University and D.C. Rine of West Virginia University for continuous motivation and help; they would also like to express their gratitude to Prof. H. Ehrig and Prof. G. Hotz for comments and references and to Prof. M. A. Arbib and Prof. J. A. Goguen for comments on a previous related draft.

* The first author's work has been financed by a Research Scholarship of the Spanish Ministry of Education and Science.

1.- SEMIMODULES AND FUZZY RELATIONS

Let A be a commutative semiring. We define the category $S\,Mod_A$ of semimodules on A with objects pairs (M, α), where M is a commutative monoid and $\alpha : A \longrightarrow End(M)$ a semiring homomorphism, and morphisms linear functions defined in the natural way. Let A( ) be the left adjoint of the forgetful functor $U: S\,Mod_A \longrightarrow Set$; then the category F_A of finite dimensional semimodules is defined with

$$|F_A| = |Fin\ Set| \quad \text{and} \quad F_A(X,Y) = S\,Mod_A\,(A(X)\,,\,A(Y))$$

If A happens to be a distributive lattice L , then F_L is isomorphic to the category Fuzz Rel_L of finite L-fuzzy relations in the sense of Goguen (2). The isomorphism generalizes the one between matrices and fixed-basis finite dimensional modules, because an L-fuzzy relation $X \xrightarrow{\varphi} Y$ is a map $\varphi : X \times Y \longrightarrow L$ that is an $X \times Y$ matrix on L. The compositions with $Y \xrightarrow{\psi} Z$ is: $\psi\varphi\,(x, z) = \bigvee_{y \in Y} (\varphi(x, y) \wedge \psi(y, z))$. Evidently we can generalize Goguen's L-fuzzy relations to any semiring A. A tensor product $\otimes$ can be defined in Fuzz Rel_A by: $X \otimes Y = X \times Y$,

$$\varphi_1 \otimes \varphi_2\,(a_1, a_2, b_1, b_2) = \varphi_1(a_1, b_1) \cdot \varphi_2(a_2, b_2)$$

We can speak about the category of sets and discrete A-fuzzy relations, isomorphic to the category of free semimodules FS Mod_A. The forgetful functor U: FS $Mod_A \rightarrow$ Set correspond to the discrete A-fuzzy "covariant power". S Mod_A, FS Mod_A and F_A have tensor products, and it is preserved in the above isomorphisms. They are closed symmetric monoidal, semiadditive, and S Mod_A has kernels and cokernels, so it is presemiabelian. The monoidal product is <u>strict</u> (3) for F_A. If A = 2, the two element boolean algebra, we recapture the standard relations Fin Rel = Fuzz Rel_2. If A is a ring, F_A are the finite dimensional modules. If A = 0,1 , we obtain Zadeh's fuzzy relations. All categories Fuzz Rel_A contain the category Fin Set of finite sets as a monoidal subcategory.

2.- TENSORS

We define for each F_A an isomorphic category T_A with the same objects and morphisms $T_A\,(X, Y) = (A\,(X))' \otimes A(Y)$ where $(A\,(X))'$ is the conjugate dual semimodule of A (X) . The <u>composition</u> in T_A relates the <u>contracted tensor product</u> of tensors and the <u>monoidal product</u>, the <u>uncontracted one</u>. If we call Fin PD = partial functions, Fin ND = non-deterministic functions, Fin Stoch stochastic functions and MT Markov matrices, and if for each category C with objects finite sets we call C^*

the full monoidal subcategory with objects $N^*(n_1 \dots n_k = [n_1] \times \dots \times [n_k])$ we obtain then the diagram:

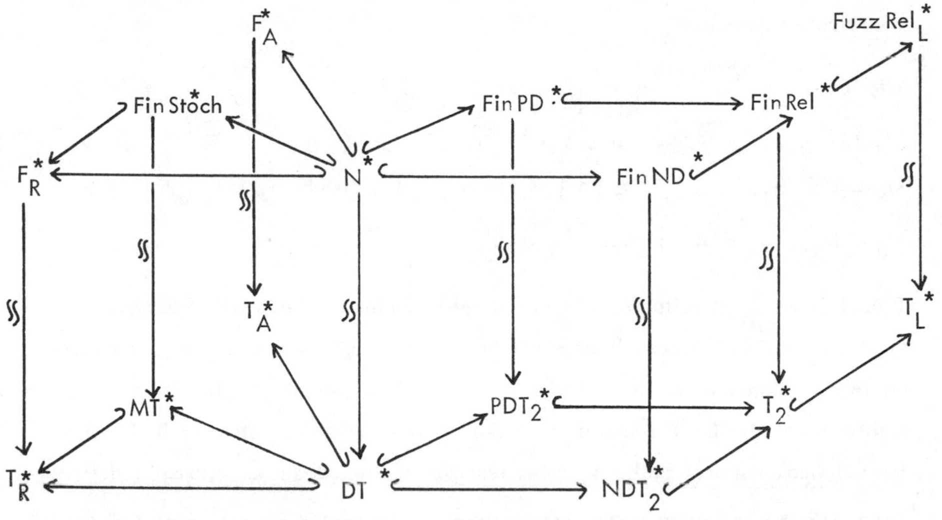

(T_R^* = real tensors, MT^* = markov tensors, T_2^* = boolean tensors, T_L^* = tensors on a lattice L). In general, morphisms in T_A^* are tensors of $A(n_1)' \otimes \dots \otimes A(n_\nu)' \otimes A(m_1) \otimes \dots \otimes A(m_\mu)$. If $n_1 = \dots = n = n_o$ (for example in the case of switching networks) we obtain the full subcategory T^*_{A,n_o} of ν-covariant, μ-contravariant tensors on $A(n_o)$ ($\nu, \mu \in N$). We obtain in this manner a tensor calculus in all these categories which is useful for dealing with coordinate expressions.

3.- AUTOMATA AND SWITCHING NETWORKS

The diagram in the preceding section and the possible generalizations of it show how all the categories for machines mentioned at the beginning can be considered as semimodule categories or at least as monoidal subcategories of them. We can then deal in this framework with all automata situations as simulations, homomorphisms, series, parallel and cascade compositions, minimal realization problems, etc. In particular following the isomorphisms down in the diagram in last section we can obtain the corresponding tensor expressions (5) . For example if we define, as in Budach-Hoehnke (9), the series composition of two Mealy machines in F_A

$(X \otimes S \xrightarrow{f} S;\ X \otimes S \xrightarrow{\varphi} Y);\ (Y \otimes T \xrightarrow{\bar{f}} T;\ Y \otimes T \xrightarrow{\bar{\varphi}} Z)$ as the machine

$$(X \otimes S \otimes T \xrightarrow{\bar{\bar{f}} = (S \otimes \bar{f})\,[(f \otimes \varphi)\, d \otimes T]} S \otimes T;\ X \otimes S \otimes T \xrightarrow{\bar{\bar{\varphi}} = \bar{\varphi}(\varphi \otimes T)} Z)$$

where d is the image by inclusion functor of the corresponding diagonal in Fin Set , and if we restrict ourselves to the $*$ situation and take $X = [n_1] \times \ldots \times [n_\nu]$ denoted by n_ν ; $S = a_\alpha$, $Y = m_\mu$, $T = b_\beta$, $Z = r_\rho$, we obtain the coordinate expression

$$\bar{\bar{F}}{}^{u_\alpha v_\beta}_{i_\nu i_\alpha k_\beta} = \sum_{w_\mu \in m_\mu} F^{u_\alpha}_{i_\nu i_\alpha} \Phi^{w_\mu}_{i_\nu i_\alpha} \bar{F}{}^{v_\beta}_{w_\mu k_\beta} \quad ; \quad \bar{\bar{\Phi}}{}^{u_\rho}_{i_\nu i_\alpha k_\beta} = \sum_{w_\mu \in m_\mu} \Phi^{w_\mu}_{i_\nu i_\alpha} \bar{\Phi}{}^{u_\rho}_{w_\mu k_\beta}$$

with each $(i_1, \ldots, i_\nu) \in [n_1] \times \ldots \times [n_\nu]$ (denoted $i_\nu \in n_\nu$), $i_\alpha \in a_\alpha$; $k_\beta \in b_\beta$; $u_\alpha \in a_\alpha$; $v_\beta \in b_\beta$; $u_\rho \in r_\rho$.

Though S Mod_A is suitable for minimal realizations in the sense of Goguen (6), FS Mod_A is not, because it does not have canonical cofactorizations. However, having in mind the results of Ehrig et al.(12) it is pseudoclosed, i.e. there is a right adjoint (U) for the inclusion of Set, and so minimal automata in FS Mod_A form a weak minimal realizing subsystematic of the adequate systematic defined there (12) for automata in the pseudoclosed case, and the more restricted one of reduced automata is furthermore a reduced subsystematic. This holds also for automata in the monoidal subcategories here considered since F_A is pseudoclosed on Fin Set, stochastic functions and discrete non-deterministic functions are pseudo closed on Set, and PD is suitable. The categories of Mealey (or Moore) automata for FS Mod_A and F_A; Aut, Aut (I), Aut (I,O), Medw, Medw (S), Medw (I) (where I,O,S are fixed input, output and state objects and Medw stands for outpuless automata), have products (except Aut (I,O) and Medw (S) for FS Mod_A and F_A), coproducts and free objects, and they are finitely complete for S Mod_A.

Pfender generalizes in his work (1) the framework of categories with finite products used by Lawvere in his thesis (7) on Algebraic Theories to the case of S-monoidal categories, where S is a fixed class of permitted substitutions ranked in a hierarchy going from strict monoidal to finite products. (In fact Pfender ranks from premonoidal to finite products and then he needs "beklammerte" natural numbers $\tilde{N}$ instead of N, but here this is not necessary). So we have Symmetric strict monoidal (Fin Stoch, F_A, ...), Terminal strict monoidal (for example Fin ND), Diagonal strict monoidal (Fin PD), Substitution strict monoidal (all substitutions permitted) and finally $\prod$ strict monoidal: strict finite products. In this manner Lawvere's theories are generalized to S-theories on some permitted substitutions category S . Free theories $F_S(\Omega)$ are defined in a similar way, and the category of T-algebras for some S-theory T and some S-monoidal category B is defined. It is particularly interesting for Switching theory because we can deal not only with deterministic but with tolerant (in the sense of Dal Cin (8)) and stochastic (combinatorial or sequential) switching networks. Given an Ω-algebra in an S-monoidal category B ; $\xi : F_S(\Omega) \longrightarrow B$, we define the corresponding S-theory of (combinatorial) switching networks as the theory having as arrows $(w, \xi(w)) : n \longrightarrow m$ for $w : n \longrightarrow m$ in $F_S(\Omega)$; intuitively we have the wires and gates pictured by w and the "calculated function" being $\xi(w) : \otimes^n \xi(1) \longrightarrow \otimes^m \xi(1)$. If B is F_A , then $\xi(1) \simeq A(n_o)$ for some $n_o \in N$ and the $\xi(w)$'s are morphisms in the S-theory F^*_{A,n_o} of tensor powers of $A(n_o)$ and by the isomorphism they are tensors of T^*_{A,n_o} . We obtain these tensors by series (contracted tensor product) and parallel (uncontracted) composition if we know the tensors for gates in Ω and for substitutions in S. The same situation holds for B a monoidal subcategory of F_A. As PD is Diagonal-monoidal, series ($\circ$) and parallel ($\otimes$) can be defined as partial operators, and so, given a suitable set of axioms, we obtain a certain composition algebra. In this algebra $\Omega \cup S$ is a generating system for $F_S(\Omega)$, and hence our tensor evaluation just arises by the natural epimorphism from the free composition algebra generated by $\Omega \cup S$. (In the sense of Hotz (10) we would construct the free X-category F(T, (1)) generated by the set T of tensors of given gates and of permitted substitutions) .

REFERENCES

1) PFENDER, M., "Universelle Algebra in monoidalen Kategorien", Technische Universität Berlin, July 1973.

2) GOGUEN, J.A., "L-Fuzzy Sets", Journ. Math. Anal. Appl. (18), 145-174 (1967).

3) MAC LANE, S., "Categories for the working Mathematician", Springer Verlag, New York, (1971).

4) MITCHELL, B., "Theory of Categories", Academic Press, New York (1965).

5) MESEGUER, J.,SOLS, I., "On a Categorical Tensor Calculus for Automata", Zaragoza University, September 1973, (unpublished paper).

6) GOGUEN, J.A., "Minimal Realization of Machines in Closed Categories", Bull AMS (78), 777-783, (1972).

7) LAWVERE, F.W., "Functorial Semantics on Algebraic Theories", Proc. Nat. Acad. Sci. USA (50), 869-873 (1963).

8) DAL CIN, M., "Fuzzy-State Automata, their Stability and Fault-Tolerance". Institute Information Sciences, University of Tübingen.

9) BUDACH, L., HOEHNKE, H.J., "Uber eine einheitliche Begründung der Automatentheorie" , Seminar Report 1. HU-Berlin. (1969-70).

10) McCULLOCK, MORENO-DIAZ , R., "On a Calculus for Triadas", Neural Networks (78-86), E.R. Caianiello (ed.), Springer Verlag, Berlin (1968).

11) ZADEH, L.A., " Fuzzy Sets", Inf. Control, (8), 338-353, (1965).

12) EHRIG, H., KIERMEIER, K.D. , KREOWSKI, H.J., KUHNEL, W., "Systematisierung der Automaten Theorie", T.U.-Berlin, (seminar 1972-73) .

13) MARTINEZ, P., MESEGUER,J., "Probabilistic Automata as Bilinear Automata", unpublished paper, April 1973.

14) MESEGUER, J., SOLS, I., 'Categorical Tensor Representation of Deterministic, Relational and Probabilistic Finite Functions" (to appear in Actas de las II Jornadas Matemàticas Hispano-Lusitanas, Madrid 1973).

15) HOTZ, G., 'Eine Algebraisierung des Syntheseproblems von Schaltkreisen " I and II, EIK (1), 185-206,209-231, (1965).

16) POYATOS, F. "Introducción a la Teoría de Semimodulos", dissertation, University of Madrid., (1967).

REPRESENTATION OF A CLASS OF NONLINEAR SYSTEMS

E. Turan Onat
Engineering and Applied Science

James Geary
Engineering and Applied Science

Yale University
New Haven, Conn. 06520, U.S.A.

The class C of systems considered in this paper arises in the study of nonlinear hereditary behavior of materials* [1]. It is defined as follows: The input and output sets X and Y are certain subsets of the linear space L composed of real valued (3x3) matrix histories on the time interval $[0,\infty)$. The system mapping F is causal and uniformly continuous in the norm chosen for L. Moreover, F is subject to the invariance requirement

$$F(Qu) = QF(u)Q^T \ , \ u \text{ and } Q \in X \tag{1}$$

where Q is a history of the matrix of proper orthogonal transformation in R_3 (Thus $Q(\tau)\varepsilon 0^+(3)$) and Qu represents the product $Q(\tau)u(\tau)$ on $\tau\varepsilon[0,\infty)$.

The purpose of the paper is to show that a system in the class C admits a state variable representation and to establish the main features of this representation. Of particular interest is the additional structure that the invariance requirement (1) endows upon the state space.

---

*In a hereditary material the current local stress tensor y(t) depends on the history of the tensor of local deformation gradients u (i.e. $u(\tau)$ on $\tau \leq t$). Viscoelastic and elastic-plastic solids are examples of such materials. In the study of distributions of stress and strain within a deforming inelastic body one needs an explicit representation of this history dependence (cf. [2]).

We now define C more precisely. We let L denote the linear space of all (3x3) matrix histories on $[0,\infty)$ whose components are real continous and bounded functions of time and addition and scalar multiplication in L is defined componentwise. We use the following norm for L:

$$||u|| = \sup_{\tau} \left(\text{tr } u(\tau)u^T(\tau) \right)^{\frac{1}{2}}, \quad \tau\varepsilon[0,\infty); \; u\varepsilon L \tag{2}$$

The input set X is composed of elements of L that satisfy the restrictions

$$u(0) = I \quad \text{and} \quad \det u(\tau) > 0, \quad \tau\varepsilon[0,\infty) \tag{3}$$

The output set Y is also a subset of L. It has the following properties:

$$y(\tau) = y^T(\tau) \qquad \tau\varepsilon[0,\infty), \; y\varepsilon Y \subset L$$

and

$$||y|| < M \qquad y \; \varepsilon \; Y \tag{4}$$

where M is a positive number. Thus outputs are symmetric matrix histories* and Y is a bounded subset of L.

A system in C is a mapping F from X to Y

$$y = F(u) \qquad u\varepsilon X, \; y\varepsilon Y \tag{5}$$

such that F is causal and uniformly continuous in the norm (2).

Furthermore, F is assumed to possess the invariance requirement (1). This requirement arises in mechanics of materials from the observation that a simultaneous rotation of the deformed body and reference system must leave the stress components unaltered (cf. [2]).

In order to give a more explicit statement of (1) consider an input u that gives rise to the output y . Next consider the input

$$Q(\tau)u(\tau) \qquad \tau\varepsilon[0,\infty)$$

where $Q(\tau)\varepsilon 0^+(3)$ = the group of proper orthogonal transformations in R_3.

The invariance requirement (1) states that the composite input Qu must give rise to the output

$$Q(\tau)y(\tau)Q^T(\tau) \quad \text{on} \quad [0,\infty)$$

---

*In terms of mechanics of materials an input is a history of the tensor of deformation gradients and an output is a history of the stress tensor.

We now give a summary of the ideas presented in the paper.

We define the notion of state as in [3] and [4] in terms of input-output pairs. This definition induces an equivalence relation on the set of input segments on [0,T] where T is a fixed time. The quotient set Σ defined by the equivalence relation is referred to as the state set. In other words Σ is the "minimal" state set constructed by Nerode equivalence.

The set Σ is then made a metric space (cf. [3]) by introducing a notion of distance based on system response on the time interval $[T,\infty)$. It is then shown that Σ is separable. Thus it is possible to speak of the dimension of Σ.

We next consider the implications of the invariance requirement (1) on Σ. Let the input segment $u_{[0,T]}$ give rise to the state S in Σ. Next consider the composite input segment

$$R(\tau)u(\tau) \quad \text{on} \quad [0,T] \tag{6}$$

where $R \varepsilon Y$ is a history of proper orthogonal transformations and

$$R(T) = Q \tag{7}$$

It is shown that the state created by (6) depends only on Q and the state S created by $u_{[0,T]}$. Thus the composite input (6) moves the state point S to a point $P_Q S \varepsilon \Sigma$ and thereby creates a transformation P_Q on Σ. It is then seen that the set of all P_Q, $Q \varepsilon 0^+(3)$ is a transformation group G on Σ.

We next show, using the work of Mostow [5], [6] that when Σ is finite dimensional and when Σ has a finite number of nonconjugate isotropy sub-groups under G (a technical restriction satisfied automatically in a "large" part of Σ) the state space Σ can be embedded in a finite dimensional euclidean space R_n in such a way that the image of G in R_n is a group of linear orthogonal transformations in R_n. This result enables us to use the theory of group representations to show that the state vector S in R_n can be expressed uniquely as a sum of irreducible tensors:

$$S = q_1 + \cdots + q_m \tag{8}$$

where each q_i defines through its components an element in an invariant subspace of R_n. Therefore,

$$P_Q S = P_Q\, q_1 + \cdots + P_Q\, q_m \tag{9}$$

where $P_Q\, q_i$ has the meaning of an ordinary tensor transformation appropriate to the <u>rank</u> of q_i.

We finally show that the representation of a "smooth" system in C has the form

$$\begin{aligned} y(t) &= f(S(t)) \\ & \qquad\qquad\qquad S \varepsilon R_n \\ \dot{S}(t) &= g(S(t),V) + T(\Omega)S(t) \end{aligned} \tag{10}$$

The first equation in (10) asserts that the present value of the output depends on the present state. The second equation provides a system of differential equations for determining the state of the system. In this equation V and Ω are the symmetric and antisymmetric parts of the matrix

$$\frac{d}{d\tau}\left(u(\tau)\; u^{-1}(t)\right),\ \tau = t$$

and they measure, respectively, the rate of deformation and rotation of a material element. $T(\Omega)$ is a well-defined linear operator in R_n that depends linearly upon Ω. Moreover the functions f and g are subject to the invariance requirements

$$\left.\begin{aligned} f(P_Q S) &= Qf(S)Q^T \\ g(P_Q S, QVQ^T) &= P_Q g(S,V) \end{aligned}\right\} \quad \text{for all } Q \varepsilon 0^+(3)$$

REFERENCES

1. Geary, J.A. and Onat, E.T., "Representation of Nonlinear Hereditary Behavior," Oak Ridge National Laboratory Technical Report, ORNL-TM-4525, June 1974.

2. Truesdell, C. and W. Noll, "The Non-linear Field Theories of Mechanics," Handbuch der Physik III, 3, Springer Verlag (1965).

3. Onat, E.T., "The Notion of State and its Implications in Thermodynamics of Inelastic Solids," pp. 292-314 in Proc. of IUTAM Sym. Vienna (1966), Springer (1968).

4. Onat, E. T., "Description of the Mechanical Behavior of Inelastic Solids," Proc. 5th U.S. Nat. Cong. Appl. Mech. (1966) pp. 421-434.

5. Mostow, G.D., "Equivariant Embeddings in Euclidean Space," Ann. of Math. 65, 3, pp. 432-446 (1957).

6. Mostow, G.D., "On a Conjecture of Montgomery," Ann. of Math. 65, 3, pp. 513-516 (1957).

DUALS OF INPUT/OUTPUT MAPS

J. Rissanen
Institutionen för systemteknik

Linköping University
S-581 83 Linköping Sweden

B. Wyman
Mathematics Department

The Ohio State University
Columbus, Ohio 43210

1. Introduction

Consider a canonical k-linear constant dynamical system $\Sigma = (U, Y, X, G, H, F)$ and its input/output map f_Σ, [1]:

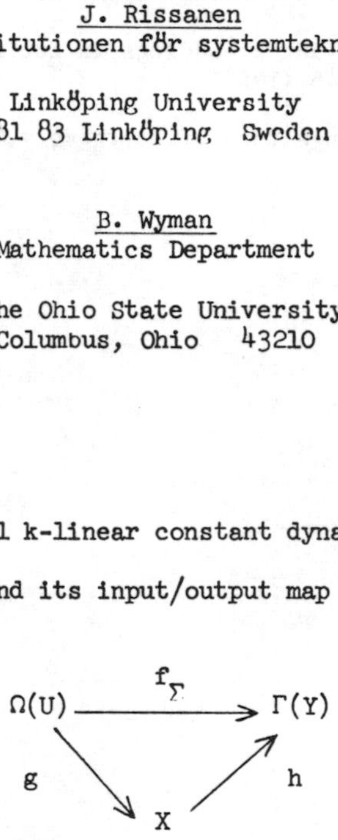

(1.1)

Here, k is a field, and U, Y and X finite dimensional k-linear spaces. $\Omega(U) = U[z] = \{ \sum_{i=-n}^{0} u_i z^{-i} | u_i \in U, \text{ some } n \geq 0\}$ is the linear space of all input strings, and $\Gamma(Y) = DY[[D]] = \{ \sum_{i=1}^{\infty} y_i D^i | y_i \in Y\}$ with $D = z^{-1}$ is the linear space of all output strings. The spaces $\Omega(U)$, $\Gamma(Y)$, and X admit a k[z]-module/structure so that the maps in (1.1) become module maps, [1].

According to a duality theory proposed by Kalman in [2, Section 10], the linear space dual of diagram (1.1) yields the so-called linear constant co-input/co-output map f_Σ^* :

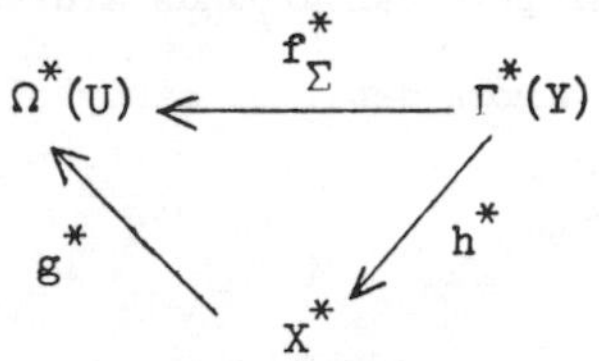

(1.2)

By starting at the other end, the realization Σ admits a dual or a corealization $\Sigma^* = (Y^*, U^*, X^*, H^*, G^*, F^*)$, [2]. The corresponding spaces of coinput strings and co-output strings, respectively are $\bar{\Omega}(Y^*) = DY^*[D]$ and $\bar{\Gamma}(U^*) = U^*[[z]]$, where the bars effect a switch $z \leftrightarrow D$, and hence the direction of time. The coinput/co-output map of Σ^* is then described by:

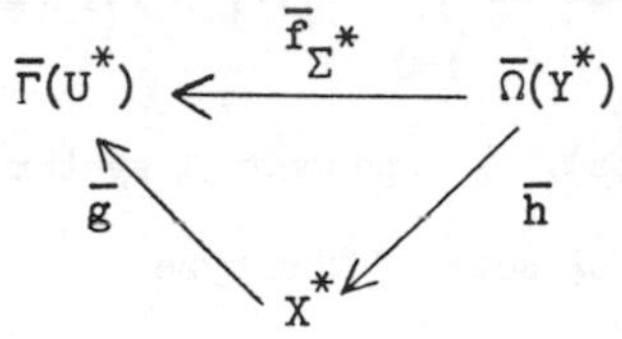

(1.3)

Kalman's main result states that the diagrams (1.2) and (1.3) are essentially the same and the canonical corealization induced by (1.2) is isomorphic to Σ^* .

Natural as such a development appears there is a serious difficulty due to the fact that $\Gamma^*(Y)$ is not a valid input string space for Σ^*; i.e. $\Gamma^*(Y)$ is not a space of the type $\bar{\Omega}(Y^*)$. Hence, a linear constant coinput/co-output map cannot be defined as f_Σ^* in (1.2), and the induced corealization does not exist.

It appears that a satisfactory duality theory cannot be developed within the "pure" algebraic apparatus which hitherto has served so well in the theory of dynamic systems. The space $\Gamma^*(Y)$ is too big and to reduce it, it seems to be necessary to introduce topologies and perform the dualization process above with respect to topological duals. Fortunately this can be done in a natural manner with topologies constructed from "scratch". We should point out that despite the module structures in (1.1) the duals needed on physical grounds are vector space duals rather that module duals. A more detailed study of the present results appears in [3].

2. Main Results

Let k have the discrete topology (even if k is the field of real numbers). Every finite dimensional k-linear space $V = k^n$ is given the product topology, which is then discrete. Further, $V[z]$, which is isomorphic to the direct sum $\bigoplus_{i=0}^{\infty} \{V_i | V_i = V\}$, is given the direct sum topology. This topology is discrete since each V_i is. Finally, $V[[z]]$, isomorphic to $\prod_{i=0}^{\infty} \{V_i | V_i = V\}$, is given the product topology (no longer discrete). A fundamental system of neighbourhoods of zero for $V[[z]]$ consists of sets of the type

$$\mathfrak{G}^n = \{ \sum_{j=n}^{\infty} v_j z^j | v_j \in V\}$$

From now on we denote by an $*$ the topological dualization operation (functor). Then for any $\varphi \in (V[[z]])^*$ the set $\varphi^{-1}(0)$ is open (since $\{0\}$ is open in k). So it must contain some $\mathfrak{G}^n$. Consider the sets:

$$\Phi_n = \{\varphi \in (DV[[D]])^* | \varphi/\mathfrak{G}^n = 0\} ,$$

where φ/A denotes the restriction of φ to A. With the collection $\{\Phi_n\}$ as a fundamental system of neighbourhoods of zero in $(V[[z]])^*$ we get a topology analogous to the usual "compact-open" topology in the theory of topological groups.

Similarly, we define topologies for $V[D]$, $V[[D]]$, and their duals. As $\Gamma(V) = DV[[D]]$ is a subset (open) of $V[[D]]$ the relative topology gives a topology for $\Gamma(V)$.

We have the following theorem:

<u>Theorem</u>. The maps

$$\alpha: \ V[z] \to (V^*[[z]])^* ,$$

$$\alpha(\omega): \ \varphi \to (\omega, \varphi) = \sum_{i=0}^{\infty} \varphi_i(v_i) ,$$

$$\beta: \ V^*[[z]] \to (V[z])^* ,$$

$$\beta(\omega): \ \varphi \to (\omega, \varphi) ,$$

are topological isomorphisms, where $(V[z])^*$ is given the obvious topology induced by the isomorphism β.

By this theorem we may identify $\Gamma^*(Y) = (DY[[D]])^*$ with the space $\bar{\Omega}(Y^*) = DY^*[D]$ of coinput strings, and $\Omega^*(U) = (U[z])^*$ with the space $U^*[[z]] = \bar{\Gamma}(U^*)$ of co-output strings. This way the diagram (1.2) defines a correct coinput/co-output map f_Σ^* with an induced canonical corealization. All told, we have the commutative diagram

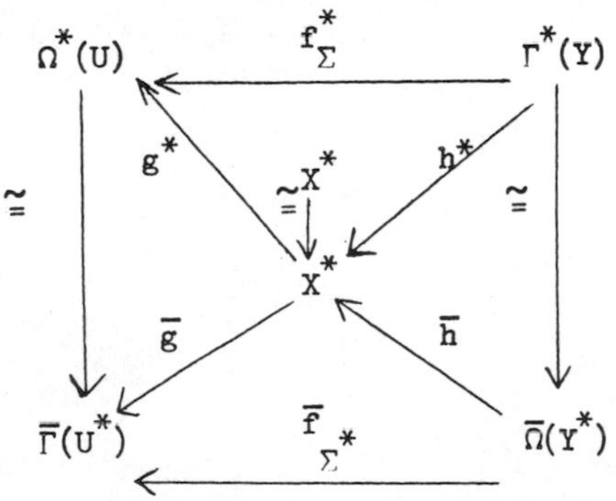

where the isomorphisms are natural.

From this Kalman's second duality theorem follows, namely, that Σ is completely reachable, observable, canonical if and only if Σ^* is completely observable, reachable, canonical; respectively.

References

[1] R. E. Kalman, P. L. Falb, M. A. Arbib, Topics in Mathematical System Theory, New York, 1969.

1. R. E. Kalman, P. L. Falb and M. A. Arbib, Topics in Mathematical System Theory, McGraw-Hill, New York, 1969.

2. R. E. Kalman, Lectures on Controllability and Observability, Centro Internationale Matematico Estivo, Bologna, Italy, 1968.

3. B. Wyman, J. Rissanen, Duals of Linear Systems and their Input/Output Maps, to appear.

Proceedings of the First International Symposium: Category Theory Applied to Computation and Control, published by the Mathematics Department and the Department of Computer and Information Science, University of Massachusetts at Amherst, 1974.

AN ALGEBRAIC FORMULATION OF THE CHOMSKY HIERARCHY

Mitchell Wand
Computer Science Department

Indiana University
Bloomington, Indiana 47401, U. S. A.

In the classic paper [2], Chomsky introduced a hierarchy of types of grammars: regular, context-free, context-sensitive, phrase-structure. This was an *ad hoc* list. More recently, the results of [1] and [8] have suggested that there is a deep algebraic relationship between the regular and context-free sets, and that a similar relationship holds between the context-free and indexed languages [6]. Thus the indexed languages may be taken as the natural next step in the hierarchy.* In this paper we define a sequence of operations E_k on theories such that for a certain theory T, the constants in $E_k(T)$ for $k = 0, 1, 2$ are precisely the regular, context-free, and indexed sets. This proves the conjecture of [9, p. 59] and [10, pp. 182-183].

---

*This view is held even by authors not normally associated with algebraic techniques, e. g. [4].

1. Definitions

We will use the properties of Cartesian closed categories, and the results of [5] in particular. Let CL denote the Cartesian closed category of small complete lattices with morphisms continuous over nonvoid directed chains. Let the usual isomorphism $\underline{CL}(L \times M, N) \to \underline{CL}(L, N^M)$ be denoted φ.

Let 1 be the singleton set, as usual. Define sets T_k of strings over the alphabet $\{0, 1, x, \to\}$ as follows:

(i) $0, 1 \in T_0$

(ii) if $t, u \in T_k$ then $*tu \in T_k$

(iii) if $t \in T_k$, then $t \in T_{k+1}$

(iv) if $t, u \in T_k$, then $\to tu \in T_{k+1}$

(v) nothing else.

Note $\cup\{T_k | k \in \omega\}$ is the set of objects of the free Cartesian closed category generated by 1, in which 0 is the terminal object and * and → have their usual meanings. Let $\underline{I}_k$ denote the full subcategory with objects T_k.

Definition: A k-theory is a functor $F : \underline{I}_k \to \underline{CL}$ preserving product and exponentiation.

Any algebraic theory has a unique extension to a k-theory $T_k(X)$, which may be built "from below" i.e., without considering objects of higher type. We often identify k-theories with their images in CL.

Let $Y \in \underline{CL}(M^M, M)$ be the morphism sending each continuous function $M \to M$ to its least fixed point. Let $\underline{A}$ be a subcategory of CL, closed under finite products. We say $\underline{A}$ is iteration-closed iff for every $g \in \underline{A}(L \times M, M)$, the morphism $Y.\varphi(g)$ belongs to $\underline{A}(L, M)$. If $\underline{A}$ is a subcategory of CL with finite products, then $\underline{A}$ has an iteration-closure $\mu(\underline{A})$.

Definition: A k-system is the iteration-closure of a k-theory

Thus every theory X has a unique extension to a k-system $\mu(T_k(X))$. We call this system $E_k(X)$.

2. Results

Theorem 1. Let $\underline{A}$ be any subcategory of $\underline{CL}$ with finite products. Then every morphism in $\mu(\underline{A})$ is of the form $E.Y.\varphi(\Pi g_i)$, where E is a projection, Y is the fixed point morphism, and each g_i is a morphism in $\underline{A}$.

This is a straightforward application of the normal form theorem for μclones [10].

Theorem 2. (Chomsky theorem) Let $k>0$, let $L \in \underline{CL}$ and for $n \geqslant 0$ let Ω_n be a set of morphisms $L^n \rightarrow L$. Let T be the theory generated by $\Omega = \coprod \Omega_n$. Then every morphism in $E_k(T)$ is of the form $E.Y.\varphi(\Pi g_i)$, where E and Y are as before, and each g_i is of the form $t_i.E_i$ where E_i is a product of projections and t_i is either

(i) $\varphi(u)$ where $u \in \Omega$

(ii) $\underline{comp} \in \underline{CL}([x \rightarrow y] * [y \rightarrow z], [x \rightarrow z])$

(iii) $\underline{eval} \in \underline{CL}([x \rightarrow y] * x, y)$

(iv) $\underline{\varphi} \in \underline{CL}([x*y \rightarrow z], [x \rightarrow [y \rightarrow z]])$

Sketch of proof. We first apply theorem 1. We then show that, without loss of generality, the g_i may be taken to be of the form $E_i.\varphi(u_i)$. We then apply the cut-elimination lemma of [9] to the u_i. This sufficiently simplifies the u_i so that a case analysis becomes feasible. We may then transform each u_i into the proper form.

Let V be a countable set (of terminal symbols), and let V* be the set of strings over V. Let $L = P(V^*)$. Let $\Omega_0 = \{\{e\}\}$ (the empty string), $\Omega_1 = \{\lambda S[aS] \mid a \in V\}$ and let $\Omega_2 = \{\lambda ST[SUT]\}$. Let T be the theory generated by Ω, and $S_k = E_k(T)$. Then $S_k(0, L)$ is a class of languages for each k.

Theorem 3. For $k = 0, 1, 2$, $S_k(0, L)$ is the class of regular, context-free, and indexed languages.

Sketch of proof. For each class $\mathcal{L}_k$, showing $\mathcal{L}_k \subseteq S_k(0, L)$ is tedious but routine. In the other direction $k = 0$ was shown in [10]. For $k = 1, 2$, we apply Theorem 2 in a weakened form, treating unions specially. Then no nontrivial multi-place functions develop, and we arrive at the Chomsky normal form theorem and Fischer's OI normal form theorem [3], respectively.

3. Conclusions and Open Problems

Our results confirm the suggestion that the (revised) Chomsky hierarchy is of algebraic, rather than *ad hoc*, origin. They also suggest some new problems. Of particular interest are the characterization of $S_\omega(0, L)$ and the question of strict hierarchy problem (i.e., is $S_k(0, L) \subsetneq S_{k+1}(0, L)$ for all k ?). In addition, the algebraic properties of E_k have not been studied. These become particularly relevant when L is the lattice of flow diagrams [7]: in this case we construct classes of flow diagrams which seem related to the theory of program schemes.

(Added in proof). Systems similar to S_k were studied by Turner (Doctoral Dissertation, University of London, 1973). When our base lattice T(1) is a power set, our systems become a special case of equational sets over a many-typed algebra, studied by Maibaum ("a Generalized Approach to Formal Languages", JCSS 8 (1974), 409-439).

REFERENCES

1. Brainerd, W.S., Tree Generating Regular Systems, Information and Control 14 (1969) 217-231.

2. Chomsky, N., On certain formal properties of grammars, Information and Control 2: 2 (1959), 137-167.

3. Fischer, M.J., Grammars with Macro-Like Productions, Proc 9th IEEE Conf Sw & Auto Th (1968), 131-142.

4. Greibach, S.A., Full AFL and Nested Iterated Substitution, Information and Control 16 (1970), 7-35.

5. Lambek J., Deductive Systems & Categories II, Category Theory, Homology Theory & Their Applications I (P. Hilton, ed.) Berlin: Springer-Verlag, Lecture Notes in Mathematics, Vol. 86, (1969), 76-122.

6. Rounds, W.C., Mappings and Grammars on Trees, Math Sys Th 4 (1970), 257-287.

7. Scott, D. The Lattice of Flow Diagrams, Oxford U. Comp. Lab. Rep. PRG-3 (1970).

8. Thatcher, J.W., Characterizing Derivation Trees of Context-Free Grammars through a Generalization of Finite Automata Theory, J Comp & Sys Sci 1 (1967), 317-322.

9. Wand, M. An Usual Application of Program-Proving Proc 5th ACM Symp on Th of Computing (Austin, 1973), 59-66.

10. Wand, M. Algebraic Foundations of Formal Language Theory, MIT Project MAC TR-108, Cambridge, Mass., 1973.

ON THE RECURSIVE SPECIFICATION OF DATA TYPES

Mitchell Wand
Computer Science Department

Indiana University
Bloomington, Indiana 47401, U. S. A.

The idea of defining data types recursively dates back at least to [1](*), e.g., "A list is either an atom or a pair of lists." In general, we want to find an object X such that $X = T(X)$, in this case, $X = A \cup X \times X$. Scott pointed out [2, 3, 4] that certain transformations T, such as $T(X) = X^X$, had no solutions in the category of sets, but there was a solution to the (weakened) equation $X \cong T(X)$ in the category of complete lattices. Scott also provided solutions for a number of interesting T's. Reynolds [5] pointed out that Scott's constructions could be unified. In this note, we point out that the unified construction is essentially categorical in nature: if T is an endofunctor on a certain category <u>CLP</u> related to the category of complete lattices, and T satisfies a certain continuity condition, then the equation $X \cong T(X)$ has a solution L_∞ which is canonical, i.e., if $M \cong T(M)$, then there exists a canonical morphism $L_\infty \to M$. This continues the program suggested by Scott [4].

*Of course, defining sets by induction is a much older idea, in general. Here we mean the application of this idea to computer programming.

Formally, let CL be the category of complete lattices, with morphisms those maps continuous over directed chains. Let CLP be the category whose objects are those of CL and with morphism sets $\underline{CLP}(L,M) = \underline{CL}(L,M) \times \underline{CL}(M,L)$. If $\varphi = \langle f,g \rangle \in \underline{CLP}(L,M)$, let $\varphi^+ = \langle g, f \rangle \in \underline{CLP}(M, L)$ (this notation follows [5]). Let CLR (the category of retractions) be the subcategory of CLP of morphisms $\varphi \in \underline{CLP}(L,M)$ such that $\varphi\varphi^+ \sqsubseteq 1_M$ and $\varphi^+\varphi = 1_L$, where the ordering is the natural one on $\underline{CLP}(L,M) = [L \to M] \times [M \to L]$. We say an endo-functor $T: \underline{CLP} \to \underline{CLP}$ is continuous on morphism sets iff $\bigsqcup T(\varphi_i) = T(\bigsqcup \varphi_i)$ for any directed set of morphisms. Let $\underline{\omega}$ be the free category generated by the graph $\{\omega, \{(n, n+1) | n \in \omega\}\}$.

Theorem 1: Let $F: \underline{\omega} \to \underline{CLR}$ be a functor, with $F(n) = L_n$ and $F(n \to n+1) = \theta_n = \langle f_n, g_n \rangle : L_n \to L_{n+1}$. Then colim F exists, and is of the form $L_\infty = \{(x_0, x_1, \ldots) | x_i \in L_i \ \&\ x_i = g_i(x_{i+1})\}$, under the ordering $x \leqslant y$ iff $x_i \leqslant y_i$ for all i.

The proof depends on the following lemma:

Lemma 1: Let $\langle f_{n\infty}, g_{\infty n} \rangle$ be the n-th component of the limiting cone. Then $\bigsqcup_n f_{n\infty} g_{\infty n} = 1_{L_\infty}$.

The main theorem is

Theorem 2: Let $T: \underline{CLP} \to \underline{CLP}$ be a functor continuous on the morphism sets, with $T(\varphi^+) = (T(\varphi))^+$. Let $\theta_0 : L_0 \to T(L_0)$ be a retraction. Define $F: \underline{\omega} \to \underline{CLR}$ as follows:

$F(0) = L_0$

$F(k+1) = T(F(k))$ on objects

$F(0 \to 1) = \theta_0$

$F(k+1 \to k+2) = T(F(k \to k+1))$ on morphisms

Let L_∞ denote colim F . Then $L_\infty \cong T(L_\infty)$.

Theorem 2 follows from Lemma 1 and Theorem 1 by "pure category theory," i.e. no representational details intrude. Then, using the fact that $\{1\}$ is an initial object in CLR , we get

Theorem 3: Let $L_0 = \{1\}$ in Theorem 2. If M is such that there exists a retraction $T(M) \to M$, then there exists a retraction $L_\infty \to M$ (unique in satisfying the appropriate diagram condition).

From this outline, it is apparent that Theorem 3 is the generalization of the fixed point theorem (for continuous functions on a complete lattice) to non-skeletal categories. Last, we illustrate a number of examples of this construction.

These results improve those of Reynolds in the following ways:

(i) The functorial nature of T is emphasized, as is the distinction between CLP and CLR .

(ii) The (co-) limit exists not just for the sequence $T^n(L_0)$ but for any sequence; furthermore, the colimit property is stronger than that normally attributed to L_∞.

(iii) The isomorphism $L_\infty \to T(L_\infty)$ arises from the colimit property and not ad hoc (in fact, we get the existence of mediating morphisms $L_\infty \to M$ for an entire class of M 's).

REFERENCES

[1] McCarthy, J. "A Basis for a Mathematical Theory of Computation" in Computer Programming and Formal Systems (ed. P. Braffort & D. Hershberg), Amsterdam, North Holland, 1963.

[2] Scott, D. Lattice-theoretic Models for Various Type-Free Calculi, Proceedings of the IVth International Congress for Logic, Methodology, and the Philosophy of Science, Bucharest, 1972 (to appear).

[3] ----- , Data Types As Lattices, Lecture Notes, Amsterdam, 1972.

[4] ----- , Continuous Lattices, in Toposes, Algebraic Geometry, and Logic, (ed. F. W. Lawvere) Springer Lecture Notes in Mathematics vol. 274, Berlin, 1972.

[5] Reynolds, J. C., Notes on a Lattice-Theoretical Approach to the Theory of Computation, Syracuse University, 1972.

LINEAR SYSTEMS OVER RINGS OF OPERATORS

B. F. Wyman
Mathematics Department

The Ohio State University
Columbus, Ohio 43210

1. Introduction

In a seminal paper written almost a decade ago, R. E. KALMAN [1965] presented an algebraic theory of stationary, discrete-time linear systems over an arbitrary field of scalars. An extensive literature, both theoretical and computational, has developed, and surveys and references can be found in KALMAN, FALB, and ARBIB [1969], and KALMAN [1969].

It was soon recognized that much of Kalman's work could be generalized to linear systems over rings of scalars. Detailed work over fairly general rings has recently begun: ROUCHALEAU [1972], ROUCHALEAU, WYMAN, and KALMAN [1972], ROUCHALEAU and WYMAN [1974?], and JOHNSTON [1973]. However, at least one worker in the subject (I am now willing to confess) had lingering doubts about the applicability of linear system theory over complicated rings. These doubts have been weakened by important work of E. W. KAMEN [1973] whose work on continuous-time system uses the idea of rings of distributions acting as operators. Some work of KALMAN and HAUTUS [1971] supports this point of view, and in particular their emphasis on "cofree modules" helped inspire the technical development in Section 3 below. Furthermore, I have been influenced by conversations with D. L. Elliott on linear systems over partially ordered sets and the idea of using the incidence algebra of a poset as a ring of operators (see ELLIOTT and MULLANS [1973]). The theory presented here is also closely related to recent research in category theory, systems, and automata done by Arbib, Manes, Goguen, and others. See, for example, ARBIB and MANES [1973]. It will be an important task to investigate this relationship more closely.

2. Discrete-Time Systems over Rings of Scalars

Let k be a commutative ring with 1.

Definition 2.1: A <u>discrete-time</u>, <u>stationary</u>, <u>linear</u> <u>dynamical</u> <u>system</u> <u>over</u> k ("k-system" for short) is a sextuple $\Sigma = (X,U,Y;\ F,\ G,\ H)$, where X, U, and Y are left k-modules and $F\colon\ X \to X$, $G\colon\ U \to X$, and $H\colon\ X \to Y$ are k-module maps.

We call elements of U "inputs", of X, "states", and of Y, "outputs".

The system Σ gives a commutative diagram of modules over the polynomial ring $S = k[z]$ as follows (see KALMAN, FALB, and ARBIB, Chap. X, for example):

(2.2)
$$\begin{array}{ccc} \Omega U & \xrightarrow{\ f_\Sigma\ } & \Gamma Y \\ & \searrow \quad {}_F X \quad \nearrow & \end{array}$$

Here ${}_F X$ is the state module X considered as an S-module via $z\,x = F\,x$, ΩU is the module of "past input strings" and ΓY is the module of "future output strings". The S-module map f_Σ is called the <u>input/output map</u> for Σ.

We will not be concerned with the exact definitions for Diagram (2.2) here, since they will turn out to be determined by rather weak and very natural assumptions. Specifically, we require that for each k-module U, ΩU is an S-module naturally associated to U in such a way that for every X and every k-module map $G\colon\ U \to X$ there is a unique S-module map $\tilde{G}\colon\ \Omega U \to {}_F X$. That is, for every X there is a natural isomorphism

$\mathrm{Hom}_k(U, X) \cong \mathrm{Hom}_S(\Omega U, X)$. Similarly, for each k-module Y, ΓY is an S-module such that for every X we have $\mathrm{Hom}_k(X, Y) \cong \mathrm{Hom}_S(X, \Gamma Y)$, with H corresponding to $\tilde{H}$.

In more precise language, consider the categories ((S-Mod)) of left S-modules and ((k-Mod)) of left k-modules. Denote the "forgetful functor" by Φ: ((S-Mod)) $\to$ ((k-Mod)), so that $\Phi({}_F X)$ is just the k-module X. To explicate the notion of "natural isomorphism" we assume that Ω: ((k-Mod)) $\to$ ((S-Mod)) is left adjoint to Φ, and Γ: ((k-Mod)) $\to$ ((S-Mod)) is right adjoint to Φ. This fundamental adjointness relation determines the structure of ΩU and ΓY uniquely, as well as the maps $\tilde{G}$ and $\tilde{H}$. (See MacLane [1972, pp. 77 ff] for basic facts about adjoint functors.) We will return to this example in Section 4a, but first we give a general discussion.

3. Systems over Rings of Operators

The central idea of the theory of Section 2 is that the "state-transition endomorphism" $F: X \to X$ allows us to consider X as a module over the polynomial ring $S = k[z]$. The foundations of the theory involve certain functors between the categories ((k-Mod)) and ((S-Mod)). The new idea of the present section is that $k[z]$ can be replaced by any k-algebra of operators, where a k-algebra is a (possibly non-commutative) ring S together with a ring homomorphism $\varphi: k \to S$, where φ maps k into the center of S. See MacLane [1963], p. 173. Once φ is given, then any S-module X can be considered as a k-module by $r \cdot x = \varphi(r) \cdot x$, with r in k and x in X. This induces a "forgetful functor" Φ: ((S-Mod)) $\to$ ((k-Mod)). Now comes the main definition.

Definition 3.1: Let k be a ring, and let $\varphi: k \to S$ be a k-algebra. Then an S/k-system is a 5-tuple $\Sigma = (X, U, Y; G, H)$, where X is an S-module, U and Y are k-modules, and $G: U \to \Phi X$ and $H: \Phi X \to Y$ are k-module maps.

Of course, we would like for an S/k-system to induce an input/output map, so we need an input-functor Ω and an output-functor Γ. Fortunately, these always exist!

Proposition 3.2: Let φ: $k \to S$ be a k-algebra. Then the forgetful functor Φ: ((S-Mod)) $\to$ ((k-Mod)) has a left adjoint Ω and a right adjoint Γ.

Proof. This is well-known, and can be found in Cartan-Eilenberg (1956, pp. 28 ff.) Explicitly, we can take $\Omega U = S \otimes_k U$, and $\Gamma Y = \mathrm{Hom}_k(S, Y)$. For Ω, S must be considered as an S-k bimodule, while the action of S on ΓY is given by $(sh)(s') = h(s's)$ for all h in ΓY, s and s' in S. Note the reversal!

Proposition 3.2 allows us to form the <u>i/o-diagram</u>

$$\begin{array}{ccc} \Omega U & \xrightarrow{\;f_\Sigma\;} & \Gamma Y \\ & {\scriptstyle \tilde{G}} \searrow \quad \nearrow {\scriptstyle \tilde{H}} & \\ & X & \end{array}$$

where $\tilde{G}$ and $\tilde{H}$ are the S-module maps corresponding to G and H under the adjunctions. The map $f_\Sigma = \tilde{H} \circ \tilde{G}$ is called the <u>input/output map</u> associated to Σ. S-modules of the form ΩU are called <u>induced modules</u> in the literature and those of the form ΓY are called <u>coinduced</u>. An (<u>abstract</u>) <u>i/o-map</u> is any S-module map f: $\Omega U \to \Gamma Y$, where U and Y are any k-modules. It will be important to develop a general realization theory which predicts when such an f comes from an S/k-system.

4. <u>Examples</u>

a) The classical case. Let $S = k[z]$. Then the action of z on a state module X given an endomorphism F: $X \to X$ and an S/k-system is a k-system as discussed in Section 1. For any U, $\Omega U = U \otimes_k S = U[z]$, which coincides with the usual construction. A k-module map G: $U \to \Phi X$ gives the expected map $\tilde{G}$: $U[z] \to X$ where $\tilde{G}(z^i u) = F^i G u$.

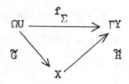

Outputs are more interesting. The general construction gives $\Gamma Y = \mathrm{Hom}_k(S, Y)$ which carries the S-module structure $(z^i h)(z^j) = h(z^{i+j})$ for all h in ΓY. If $H: \Phi X \to Y$ is a k-module map, then $\tilde{H}: X \to \Gamma Y$ is given by $\tilde{H}(x)(z^i) = H(z^i x)$. Since ΓY is uniquely determined by abstract nonsense, $\mathrm{Hom}_k(S, Y)$ must be closely related to the usual z-transform techniques, and in fact, every $h: k[z] \to Y$ corresponds to a formal power series $\sum_{i=0}^{\infty} h(z^i) \cdot z^{-i-1}$. The S-module structure on the collection of power series corresponds exactly to the one originally introduced by Kalman: $z(\sum_{i=1}^{\infty} a_i z^{-i}) = \sum_{i=1}^{\infty} a_{i+1} z^{-i}$ or "shift left and truncate the left-hand end."

b) Systems with a known monic recurrence. Suppose given a $k[z]/k$-system $\Sigma = (X, U, Y; F, G, H)$. Let $\psi(z)$ be a monic polynomial in $k[z]$ such that $\psi(F) = 0$. For example ψ could be the minimal or characteristic polynomial of F, in case k is a field. Since S is finitely generated as a k-module, this construction, together with a "change of operator ring" technique, gives a convenient and powerful approach to realization theory in the classical case.

c) Partial difference equations. Let $S = k[z_1, z_2]$ be a polynomial ring in 2 variables (although any finite number of variables will do.) In this theory, $\Omega U = U[z_1, z_2]$ and $\Gamma Y = Y[[z_1^{-1}, z_2^{-1}]]$ where appropriate truncations are used to define the S-module structure on ΓY. This example codifies a multivariable z-transform approach to partial difference equations with constant coefficients.

d) Systems over partially ordered groups. Let P be a directed pogroup (FUCHS [1963]) with neutral element e. Let $P^- = \{p \mid p \leq e\}$ and let S be the semigroup ring $k[P^-]$. The resulting theory of S/k-systems is a natural generalization of Examples (a) (take $P = Z$) and (c) (take $P = Z \oplus Z$).

e) Continuous-time systems. Let $k = R$ be the field of real numbers, and let S be the space of Schwarz distributions with compact support contained in $(-\infty, 0]$. Then S is an integral domain with convolution as multiplication. In work cited earlier, Kalman and Hautus mention the Γ functor for S/R-systems. E. W. Kamen has studied systems over rings generated by special distributions.

References

1. Arbib, M. A. and Manes, E. G. [1973a], "Machines in a Category," Tech. Machines, and Duality," Tech. Rep. 73B-1. Dept. of Computer and Inform. Sci., Univ. of Mass at Amherst.

2. Cartan, H. and Eilenberg, S. [1956], Homological Algebra, Princeton.

3. Elliott, D. L. and Mullans, R. [1973], "Linear Systems over partially ordered Sets," to be published.

4. Fuchs, L. [1963], Partially Ordered Algebraic Systems, London.

5. Johnston, R. de B. [1973], "Linear Systems over Various Rings," M.I.T. Electronic Systems Lab. Rep. ESL-R-497, (M.I.T. Dissertation).

6. Kalman, R. E. [1965], Algebraic Structure of Linear Dynamical Systems. I. The Module of Σ, Proc. Nat. Acad. Sci. (USA) 54: 1503-1508.

7. __________, [1969], Lectures on Controllability and Observability, C.I.M.E. Summer Course, Bologna, 1968. Cremonese, Rome.

8. __________, Falb, P., and Arbib, M. A. [1969], Topics in Mathematical System Theory, New York.

9. __________ and Hautus, M. [1972], Realization of Continuous-time Linear Dynamical Systems, in Ordinary Differential Equations (Ed. L. Weiss), New York.

10. Kamen, E. W. [1973], "On an Algebraic Theory of Systems Defined by Convolution Operators," to be published.

11. MacLane, S. [1963], Homology, Berlin-Heidelberg-New York.

12. __________ [1972], Categories for the Working Mathematician, Berlin-Heidelberg-New York.

13. Rouchaleau, Y. [1972], "Linear, Discrete Time, Finite Dimensional Dynamical Systems over Some classes of Commutative Rings," Dissertation, Stanford Univ.

14. __________, and Wyman, B. F. [1974?], Linear Dynamical Systems over Integral Domains, to appear in J. Comp. Sys. Sci.

15. __________, __________, and Kalman, R. E., [1972], Algebraic Structure of Linear Dynamical Systems. III. Realization Theory over a Commutative Ring, Proc. Nat. Acad. Sci. (USA), 69: 3404-3406.

THE TRICOTYLEDON THEORY OF SYSTEM DESIGN

Dr. A. Wayne Wymore
Systems & Industrial Engineering

The University of Arizona
Tucson, Arizona, 85721, USA

If scientists and engineers are to be able to attack social problems with the same validity and effectiveness with which physical problems have been attacked, then a much more sophisticated and rigorous system design methodology is needed.

Such a methodology, called the tricotyledon theory of system design, has been developed as a synthesis of: empirical and speculative approaches to system design [2], the methodology of operations research [10], classical optimization techniques [18], results in specialized branches of mathematical system theory [1,4,6,7], and the constructs of general mathematical system theory [12,13,14].

Experimental applications of the tricotyledon theory of system design to several problems in widely differing contexts [3,4,8,9,16,17] indicate that the methodology holds great promise for validity and effectiveness in attacking <u>any</u> large-scale, complex, man/machine system design problem with precision and rigor. There are <u>no</u> theoretical limitations (such as differentiability, linearity, finiteness, or discreteness) for the problems that can be attacked within the tricotyledon theory of system design.

In this short abstract the theory is only described in terms of set and system theoretic concepts as presented in Reference [12], [13], or [14];

no exploitation of the theory, either theoretical or practical is attempted here. The emphasis in the tricotyledon theory of system design is on stating a large-scale, complex, man/machine system design problem as comprehensively yet as precisely as possible, without also stating the solution, as an optimization problem, where the set over which optimization is sought is not a finite dimensional Euclidean vector space nor a set of real-valued functions, but is, essentially, a set of systems where system is a set theoretic construct.

A system design problem is a 6-tuple, $P=(S,T,\alpha,\beta,\gamma,\mathcal{D})$. The artifacts making up the 6-tuple are developed in practice as the answers to fundamental questions negotiated with the client:

What is the System Supposed to Do, Basically? The question is answered by defining an input/output specification S: An input/output specification is a 5-tuple: $P=(P,F,Q,G,\eta)$ where: P is a set not empty; $F \subset$ *FUNCTIONS(REALS,* P*)*, $F\neq\emptyset$; Q is a set not empty; $G \subset$ *FUNCTIONS(REALS,* Q*)* and $G\neq\emptyset$; $\eta\varepsilon$*FUNCTIONS(*F*,SUBSETS*(G)). The set of inputs specified by S, the set of input functions specified by S, the set of outputs specified by S, the set of output functions specified by S, the function specified by S for matching input functions with subsets of output functions, and the set of transfer functions of S are denoted, respectively, *INPUTS(S)*, *INPUTFUNCTIONS(S)*, *OUTPUTS(S)*, *OUTPUTFUNCTIONS(S)*, *matchingfunction(S)*, and *TRANSFERFUNCTIONS(S)*, and are defined, respectively, as follows: *INPUTS(S)*=P; *INPUTFUNCTIONS(S)*=F; *OUTPUTS(S)*=Q; *OUTPUTFUNCTIONS(S)*=G; *matchingfunction(S)*=η; *TRANSFERFUNCTIONS(S)*=$\{\psi:\psi\varepsilon$*FUNCTIONS(INPUTFUNCTIONS(S), OUTPUTFUNCTIONS(S))* such that for every fε*INPUTFUNCTIONS(S)*, $\psi(f)\varepsilon$(*matchingfunction*(S))(f)$\}$.

What Resources Do We Have at Our Disposal to Build the System? The question is answered by defining a technology T: A technology is a set T not empty of systems, closed under system isomorphism, that is, if $Z\varepsilon T$ and Z' is system isomorphic to Z, then $Z'\varepsilon T$.

How Is the System's Performance to be Judged? The question is answered by defining a merit ordering α over the input/output cotyledon determined by the input/output specification *S*: An input/output specification *S* is satisfied(x,ζ,T) by an assemblage *Z* if and only if: $x \in$ *STATES(Z); INPUTS(Z)=INPUTS(S); INPUTFUNCTIONS(Z)* $\supset$ *INPUTFUNCTIONS(S);* $\zeta \in$ *FUNCTIONS(STATES(Z),OUTPUTS(S));* $T \subset$ *TIMESCALE(Z),* $T \neq \emptyset$; for every $f \in$ *INPUTFUNCTIONS(S)*, there exists $g \in$ *((matchingfunction(S))*(f)) such that for every $t \in T$, ζ(((*motion(Z))*(f,t))(x))=g(t). The input/output cotyledon determined by the input/output specification *S* is denoted *IOCOTYLEDON*(*S*) and is defined as follows: *IOCOTYLEDON(S)*={(Z,x,ζ,T): *Z* is a system and *S* is satisfied(x,ζ,T) by *Z*}. The set of input/output merit orderings of the input/output specification *S* is denoted *IOMERITORDERINGS(S)* and is defined as follows: *IOMERITORDERINGS(S)=ORDERINGS(IOCOTYLEDON*(*S*)).

How Do We Judge How Well We Use Our Resources? The question is answered by defining a merit ordering over the technology cotyledon determined by a technology *T*: A system *Z* is buildable in a technology *T* if and only if there exists a coupling recipe K such that *COMPONENTS*(K) $\subset$ *T* and *Z=RESULTANT*(K). The technology cotyledon determined by the technology *T* is denoted *TECHOTYLEDON(T)* and is defined as follows: *TECHOTYLEDON(T)*= ={(K,Z): K is a coupling recipe, *COMPONENTS(K)* $\subset$ *T*, Z is a system, *Z=RESULTANT*(K)}. The set of technology merit orderings of the technology *T* is denoted *TECHMERITORDERINGS(T)* and is defined as follows: *TECHMERIT-ORDERINGS(T) = ORDERINGS*(*TECHOTYLEDON*(T)).

How Do We Resolve Conflicts Between Performance and Resource Utilization? The question is answered by defining a merit tradeoff ordering over the feasibility cotyledon determined by an input/output specification and a technology, consistent with the feasibility extensions of an input/output merit ordering and a technology merit ordering: A system *Z'* simulates(Z'',ρ,μ,θ) a system *Z* if and only if *Z''* is a subsystem of *Z'* and *Z* is an homomorphic(ρ,μ,θ) image of *Z''*. A system *Z'* simulates a system *Z* if and only if there exists Z'',ρ,μ, and θ such that *Z* is simulated(Z'',ρ,μ,θ) by Z'. A system Z is implementable in a technology T if and only if there exists a system Z' buildable in the technology T such that Z' simulates Z. The feasibility cotyledon determined by S and T is denoted *FEASIBILITYCOTYLEDON*(S,T) and is defined as follows: *FEASIBILITYCOTYLEDON*$(S,T)=\{(Z,x,\zeta,T,K,Z',Z'',\rho,\mu,\theta)$: $(Z,x,\zeta,T)\varepsilon$*IOCOTYLEDON*(S), $(K,Z')\varepsilon$*TECHOTYLEDON*(T), Z is simulated(Z'',ρ,μ,θ) by $Z'\}$. Let $u=(Z,x,\zeta,T,K,Z',Z'',\rho,\mu,\theta)$, $u\varepsilon$*FEASIBILITYCOTYLEDON*(S,T), then the input/output element in the feasible solution u, the technology element in the feasible solution u, and the system simulating model artifact of the feasible solution u, are denoted, respectively: *IOELEMENT*(u), *TECHELEMENT*(u), and *SYSTEMODEL*(u), and defined, respectively, as follows: *IOELEMENT*$(u)=(Z,x,\zeta,T)$; *TECHELEMENT*$(u)=(K,Z')$; *SYSTEMODEL*$(u)=Z'$.

Let α be a merit ordering over *IOCOTYLEDON*(S); let β be a merit ordering over *TECHOTYLEDON*(T). Then the feasibility extension of α and the feasibility extension of β are denoted, respectively, *feasibilityextension*(α), and *feasibilityextension*(β), and are defined, respectively, as follows: if $u,v\varepsilon$*FEASIBILITYCOTYLEDON*(S,T) then (*feasibilityextension*(α))$(u,v) = \alpha($*IOELEMENT*$)(u)$, *IOELEMENT*$(v))$; (*feasibilityextension*(β))(u,v) $=\beta($*TECHELEMENT*(u), (*TECHELEMENT*)$(v))$.

Let A be a set and let $\alpha,\beta,\gamma\varepsilon$*ORDERINGS*(A). Then γ is consistent with α and β if and only if: for $a,b\varepsilon A$,

$\alpha(a,b)=\beta(a,b)=1$ imply $\gamma(a,b)=1$; $\alpha(b,a)=\beta(b,a)=1$, and,
$\alpha(a,b)=0$ or $\beta(a,b)=0$, imply $\gamma(a,b)=0$.

The set of feasibility merit tradeoff orderings for the input/output specification S, the technology T, the input/output merit ordering α, and the technology merit ordering β is denoted *MERITRADEOFFORDERINGS*(S,T,α,β) and is defined as follows: *MERITRADEOFFORDERINGS*$(S,T,\alpha,\beta)=\{\gamma:\gamma\varepsilon$*ORDERINGS* (*FEASIBILITYCOTYLEDON*(S,T)), γ is consistent with *feasibilityextension*(α) and *feasibilityextension*$(\beta)\}$.

A merit ordering can be defined in terms of a figure of merit and an ordering of merit: If C is an input/output cotyledon, a technology cotyledon, or a feasibility cotyledon, then a figure of merit over C is any function defined over C with values in a non empty set. If f is a figure of merit over C and α*εORDERINGS(RANGE*(f)), then α is called an ordering of merit with respect to C and the merit ordering over C determined by f and α is denoted *meritordering*(C,f,α) and is defined as follows: if $(x,y)\varepsilon C^2$, then (*meritordering*(C,f,α))(x,y)=α(f(x),f(y)). A figure of merit is frequently defined as the expected value of a performance index: If *Z* is a system, then a performance index for *Z* is any function defined on *SYSTEMEXPERIMENTS*(*Z*) with values in a non empty set.

How Ought the System Be Tested Once It Is Built? The question is answered by defining a system test plan. Let A be a set. Let α*εORDERINGS*(A). Then the equivalence relation determined by α is denoted *equivalence*(α) and is defined as follows: if a,bεA, then (*equivalence*(α))(a,b)=1 if and only if α(a,b)=α(b,a)=1. Now let α*εEQUIVALENCERELATIONS*(A). Then the partition of A determined by α is denoted *PARTITION*(A,α) and is defined as follows: *PARTITION*(A,α)={{a':a'εA,α(a,a')=1}: aεA}. The equivalence relation α over A is finite if and only if *PARTITION*(A,α)*εFINITESUBSETS(SUBSETS*(A)). Now suppose that α,β*εEQUIVALENCERELATIONS*(A). Then β is consistent with α if and only if for every a,bεA, if α(a,b)=1, then β(a,b)=1.

Let *S* be an input/output specification; let *T* be a technology; let γε*(ORDERINGS(FEASIBILITYCOTYLEDON*(*S*,*T*)). Then the set of test decision frameworks for the *S*/*T* feasibility cotyledon, consistent with γ, is denoted *TESTDECISIONFRAMEWORKS*(*S*,*T*,γ) and is defined as follows: *TESTDECISIONFRAMEWORKS*(*S*,*T*,γ) = {δ:δ is a finite equivalence relation over *FEASIBILITYCOTYLEDON*(*S*,*T*) consistent with *equivalence*(γ)}.

If δε*TESTDECISIONFRAMEWORKS*(*S*,*T*,γ), then the set of system tests for the *S*/*T* feasibility cotyledon with the test decision framework δ is denoted *SYSTEMTESTS*(*S*,*T*,δ) and is defined as follows: *SYSTEMTESTS*(*S*,*T*,δ) ={*d*:*d* is a function such that *DOMAIN*(*d*) = *FEASIBILITYCOTYLEDON*(*S*,*T*)X *SYSTEMS* and, if (u,Z^R)ε*FEASIBILITYCOTYLEDON*(*S*,*T*)X*SYSTEMS*, then $d(u,Z^R)$ε {u *belongs to* M:Mε*PARTITION(FEASIBILITYCOTYLEDON*(*S*,*T*),δ)}X {*SYSTEMODEL*(u) *is an adequate model of* Z^R, *SYSTEMODEL*(u) *is not an adequate model of* Z^R} X {Z^R *is acceptable,* Z^R *is not acceptable*}}.

The set of system test plans for the *S*/*T* feasibility cotyledon with merit ordering γ is denoted *TESTPLANS*(S,T,γ) and is defined as follows: *TESTPLANS*(S,T,γ)={(δ,d): $\delta\varepsilon$*TESTDECISIONFRAMEWORKS*(S,T,γ), $d\varepsilon$*SYSTEMTESTS*(S,T,δ)}.

The task of defining a large-scale, complex, man/machine system design problem $P=(S,T,\alpha,\beta,\gamma,D)$ often requires an interdisciplinary team in practice. An interdisciplinary team requires, in turn, that the above definitions be translated into non-mathematical methodology. This has been done in Reference [15].

Once a system design problem $P=(S,T,\alpha,\beta,\gamma,D)$ has been defined, the objective of further effort is to find an element uε*FEASIBILITYCOTYLEDON*(S,T) optimal or, at least maximal with respect to γ:

If A is a set and $\gamma\varepsilon$*ORDERINGS*(A), and $a^*\varepsilon A$, then a^* is optimal with respect to γ if and only if for every $b\varepsilon A$, $\gamma(b,a^*)=1$. An element $a^{**}\varepsilon A$ is maximal with respect to γ if and only if for every $b\varepsilon A$, either $\gamma(b,a^{**})=1$ or $\gamma(a^{**},b)=0$.

REFERENCES

1. Arbib, M.A., Theories of Abstract Automata, Prentice-Hall, Englewood Cliffs, New Jersey, 1968.
2. Asimow, M., Introduction to Design, Prentice-Hall, Englewood Cliffs, New Jersey, 1962.
3. Blood, B.D. Jr., Toward An Optimum Programming Language for Communications Computers, Master's Thesis, The University of Arizona, Tucson, Arizona, 1973.
4. Hartmanis, J. and R.E. Stearns, Algebraic Structure Theory of Sequential Machines, Prentice-Hall, Englewood Cliffs, New Jersey, 1966.
5. Ickler, R.C., Systems Engineering Methodology Applied to the Problem of Creating a Management Organization, Master's Thesis, The University of Arizona, Tucson, Arizona, 1973.
6. Kalman, R.E., D.L. Falb, and M.A. Arbib, Topics in Mathematical System Theory, McGraw-Hill, New York, 1969.
7. Krohn, K., and J. Rhodes, "Algebraic Theory of Machines I: Prime Decomposition Theorem for Finite Semigroups and Machines," Transactions of the American Mathematical Society, Vol. 116, No. 4, April, 1965, pp. 450-464.
8. Lee, C.F., Control Systems Design and Optimization, A Rigorous Mathematical Approach by Means of a Mathematical Theory of Systems Engineering, Master's Thesis, The University of Arizona, Tucson, Arizona, 1973.
9. Rogers, J.J., Design of a System For Predicting Effects of Vegetation Manipulation on Water Yield in the Salt-Verde Basin, Ph.D. Dissertation, The University of Arizona, Tucson, Arizona, 1973.
10. Wagner, H.M., Principles of Operations Research, Prentice-Hall, Englewood Cliffs, New Jersey, 1969.
11. Wilde, D.J. and C.S. Beightler, Foundations of Optimization, Prentice-Hall, Englewood Cliffs, New Jersey, 1967.
12. Wymore, A.W., A Mathematical Theory of Systems Engineering: The Elements, John Wiley, New York, 1967.
13. Wymore, A.W., "Discrete Systems and Continuous Systems," in Advances in Mathematical Systems Theory, edited by P. Hammer, The Pennsylvania State University Press, University Park, Pennsylvania.
14. Wymore, A.W., "A Wattled Theory of Systems," in Trends in General System Theory, edited by G.J. Klir, John Wiley, New York, 1972.
15. Wymore, A.W., Systems Engineering Methodology for Interdisciplinary Teams, unpublished manuscript.
16. Zapata, R.N., A.W. Wymore and B.K. Cross, "A Systems Engineering Formulation of the Open Pit Mine Problem," Proceedings of the Eleventh International Symposium on Computer Applications in the Mineral Industry, Tucson, Arizona, 1973.
17. Zapata, R.N., and A.W. Wymore, "A Systems Engineering Formulation of the Information and Referral Problem," Proceedings of the 1973 IEEE Conference on Man, Systems, and Cybernetics, Boston, Massachusetts, November 5-7, 1973.

A WORKING BIBLIOGRAPHY FOR CATEGORICAL SYSTEM THEORISTS

compiled by

E. G. Manes
Mathematics

University of Massachusetts
Amherst, MA 01002 U.S.A.

Items for future updatings of this bibliography should be sent to E. G. Manes at the address above.

A. V. Aho ed., Currents in Computing, Prentice-Hall, 1973.

A. V. Aho and J. D. Ullman, The Theory of Parsing, Translation and Compiling, Prentice-Hall, 1972.

S. Alagić, Natural state transformations, Department of Computer and Information Science Technical Report 73B-2, Univ. of Mass., 1973.

S. Alagić, Algebriac aspects of Algol 68, Computer and Information Science Technical Report 73B-5, Univ. of Mass., 1973.

S. Alagić, Categorical theory of tree transformations, Proceedings of the First International Symposium: Category Theory Applied to Computation and Control, Univ. of Mass., 1974.

P. D'Alessandro, A. Isidori and A. Ruberti, Realization and structure theory of bilinear dynamical systems, SIAM J. Cont., to appear.

B.D.O. Anderson, Linear multivariable control systems - a survey, Proceedings of the 5th IFAC World Congress, 1972, 1-6.

B.D.O. Anderson, Algebraic properties of minimal degree spectral factors, to appear.

B.D.O. Anderson, M. A. Arbib and E. G. Manes, Finitary and infinitary conditions in categorical realization theory, to appear.

H. Appelgate, Acyclic models and resolvent functors, dissertation, Columbia University, 1965.

M. A. Arbib, Categories of (M,R)-systems, Bull. Math. Biophy. 28, 1966, 511-517.

M. A. Arbib, Automata theory and control theory: a rapprochement, Automatica 3, 1966, 161-189.

M. A. Arbib ed., Algebraic Theory of Machines, Languages and Semigroups, Academic Press, 1968.

M. A. Arbib, Theories of Abstract Automata, Prentice-Hall, 1969.

M. A. Arbib and Y. Give'on, Algebra automata I: parallel programming as a prolegomena to the categorical approach, Inf. Cont. 12, 331-345.

M. A. Arbib and E. G. Manes, Machines in a category: an expository introduction, SIAM Review, to appear.

M. A. Arbib and E. G. Manes, Foundations of system theory: decomposable machines, Automatica, to appear.

M. A. Arbib and E. G. Manes, Adjoint machines, state-behavior machines, and duality, J. Pure Appl. Alg., to appear.

M. A. Arbib and E. G. Manes, Category theory in the system sciences, Proceedings of the Tagung über Kategorien, Mathematisches Forschungsinstitut Oberwolfach, 1973, 38-40.

M. A. Arbib and E. G. Manes, Fuzzy morphisms in automata theory, Proceedings of the First International Symposium: Category Theory Applied to Computation and Control, Univ. of Mass., 1974.

M. A. Arbib and E. G. Manes, Time-varying systems, Proceedings of the First International Symposium: Category Theory Applied to Computation and Control, Univ. of Mass., 1974.

M. A. Arbib and E. G. Manes, The Categorical Imperative: Arrows, Structures, and Functors, Academic Press, to appear.

M. A. Arbib and H. P. Zeiger, On the relevance of abstract algebra to control theory, Automatica 5, 1969, 589-606.

I. Băiann and M. Marinescu, Organismic supercategories: I. Proposals for a general unitary theory of systems, Bull. Math. Biophy. 30, 1968, 625-635.

E. S. Bainbridge, A unified minimal realization theory, with duality for machines in a hyperdoctrine, dissertation, Univ. of Mich., 1972.

E. S. Bainbridge, Addressed machines and duality, Proceedings of the First International Symposium: Category Theory Applied to Computation and Control, Univ. of Mass., 1974.

J. L. Baker, Factorization of Scott-style automata, Proceedings of the First International Symposium: Category Theory Applied to Computation and Control, Univ. of Mass., 1974.

M. Barr, Coequalizers and free triples, Math. Zeit. 116, 1970, 307-322.

M. Barr, Factorizations, generators and rank, to appear.

M. Barr, Right exact functors, to appear.

F. Bartholomes and G. Hotz, Homomorphismen und Reduktionen linearer Sprachen, Lecture Notes in Operations Research and Mathematical Systems 32, 1970.

J. Beck, Triples, algebras and cohomology, dissertation, Columbia University, 1967; available from University Microfilms, Ann Arbor, Michigan.

J. Beck, Distributive laws, Lecture Notes in Mathematics 80, Springer-Verlag, 1969, 119-140.

F. S. Beckman, Categorical notions and duality in automata theory, IBM T. J. Watson Research Center Report RC 2977, 1970.

H. Bekic, Definable operations in general algebra, and the theory of automata and flowcharts, Report IBM Laboratory Vienna, 1969.

R. Bellman, Control theory, Sci. Amer., September 1964, 186-200, 272.

D. B. Benson, Syntax and semantics: a categorical view, Inf. Cont. 17, 1970, 145-160.

D. B. Benson, Semantic preserving translations, Math. Syst. Th., to appear.

D. Benson, An abstract machine theory for formal language parsers, Proceedings of the First International Symposium: Category Theory Applied to Computation and Control, Univ. of Mass., 1974.

D. B. Benson, An abstract machine theory for formal language parsers, Acta Informatica, to appear.

D. B. Benson, The basic algebraic structures in categories of derivations, Inf. Cont., to appear.

E. R. Berlekamp, Algebraic Coding Theory, McGraw-Hill, 1968.

E. Bertsch, Surjectivity of functors on grammars, Berichte des Sonderforschsungsbereichs Elektronische Sprachforschung.

L. S. Bobrow and M. A. Arbib, Discrete Mathematics: Applied Algebra for Computer and Information Science, Saunders, 1974.

W. S. Brainerd, The minimization of tree automata, Inf. Cont. 13, 1968, 484-491.

W. Brauer, Zu den Grundlagen einer Theorie topologischer sequentialler Systeme und Automaten, Habilitationschrift Universität Bonn 1970, Bericht Gesellschaft Math. Datenverarb. Bonn 31, 1970.

R. W. Brockett, Finite-Dimensional Linear Systems, Wiley, 1970.

R. W. Brockett, System theory on group manifolds and coset spaces, SIAM J. Cont. 2, 1972, 265-284.

R. W. Brockett and A. S. Willsky, Finite-state homomorphic sequential machines, IEEE Trans. Aut. Cont. AC-17, 1973, 483-490.

R. W. Brockett and A. S. Willsky, Some structural properties of automata defined on groups, Proceedings of the First International Symposium: Category Theory Applied to Computation and Control, Univ. of Mass., 1974.

J. A. Brzozowski, Regular expressions for linear sequential circuits, IEEE Trans. EC-14, 1965, 148-156.

I. Bucur and A. Deleanu, Theory of Categories, Academic Press, 1969.

L. Budach, Automata in additive categories with applications to stochastic linear automata, Proceedings of the First International Symposium: Category Theory Applied to Computation and Control, Univ. of Mass., 1974.

L. Budach, Sequentielle Gruppen, Mitt. der Math. Gesell. der DDR 23, 1973, 12-23.

L. Budach, Group objects in the category of automata, Proceedings of the Conference on Algebraic Methods in Automata Theory, Szeged, 1973, to appear.

L. Budach and H.-J. Hoehnke, Uber eine einheitliche Begrundung der Automatentheorie, Seminarbericht 1. Teil, Humboldt Universität, Berlin, 1969/70.

L. Budach and H.-J. Hoehnke, Automaten und Funktoren, Akademieverlag Berlin, to appear.

E. Burroni, Algèbres relatives à une loi distributive, Comp. R. Acad. Sc. Paris, t. 276 (5 fév. 1973), Ser. A, 443-446.

E. Burroni, Algèbres non détérministique, Proceedings of the Tagung über Kategorien, Mathematisches Forschungsinstitut Oberwolfach, 1973, page 9.

R. W. Burstall and J. W. Thatcher, Algebraic theory of recursive program schemes, Proceedings of the First International Symposium: Category Theory Applied to Computation and Control, Univ. of Mass., 1974.

L. A. Carlson, Realization is continuously universal, Proceedings of the First International Symposium: Category Theory Applied to Computation and Control, Univ. of Mass., 1974.

J. Carlyle, On the external probability structure of finite state channels, Inf. Cont. 7, 1964, 385-397.

N. Chomsky, On certain formal properties of grammars, Inf. Cont. 2, 1959, 136-67.

N. Chomsky, Aspects of the Theory of Syntax, M.I.T. Press, 1965.

N. Chomsky and M. P. Schutzenberger, The algebraic theory of context-free languages, in Computer Programming and Formal Systems, North-Holland, 1963, 118-161.

E. A. Coddington and N. Levison, Theory of Ordinary Differential Equations, McGraw-Hill, 1955.

M. Cohn and S. Even, Identification and minimization of linear machines, IEEE Trans. EC-14, 1965, 367-376.

P. M. Cohn, Universal Algebra, Harper and Row, 1965.

S. Comorasan and I. Băianu, Abstract representations of biological systems in supercategories, Bull. Math. Biophy. 31, 1969, 59-70.

R. Davis, Universal coalgebra and categories of transition systems, Math. Sys. Th. 4, 1969, 91-95.

W. A. Davis and J. A. Brzozowski, On the linearity of sequential machines, IEEE Trans. EC-15, 1966, 21-29.

C. A. Desoer and E. S. Kuh, Basic Control Theory, McGraw-Hill, 1970.

D. R. Deuel, Time-varying linear sequential machines, dissertation, Univ. of Cal. at Berkeley, 1967.

J. E. Doner, Tree acceptors and some of their applications, J. Comp. Sys. Sci. 4, 1970, 406-451.

E. J. Dubuc, Kan Extensions in Enriched Category Theory, Lecture Notes in Mathematics 145, Springer-Verlag, 1970.

E. J. Dubuc, Free monoids, J. Alg., to appear.

A. E. Eckberg, The role of canonical matrices in linear system theory, Eighth Annual Princeton Conference on Information Sciences and Systems, 1974, to appear.

B. Eckmann ed., Seminar on Triples and Categorical Homology Theory, Lecture notes in mathematics 80, Springer-Verlag, 1969.

H. Ehrig, Automata theory in monoidal categories, Proceedings of the Tagung über Kategorien, Mathematisches Forschungsinstitut Oberwolfach, 1972, 12-15.

H. Ehrig, K. D. Kiermeier, H.-J. Kreowski and W. Kühnel, Systematisierung der Automatentheorie, Seminarbericht, Technische Universität Berlin, Fachbereich Kybernetik, 1973.

H. Ehrig and H.-J. Kreowski, Power and initial automata in pseudoclosed categories, Proceedings of the First International Symposium: Category Theory Applied to Computation and Control, Univ. of Mass., 1974.

H. Ehrig, H.-J. Kreowski and M. Pfender, Kategorielle Theorie der Reduktion, Minimerung und Äquivalenz von Automaten, Math. Nachr., to appear.

H. Ehrig and M. Pfender, Kategorien und Automaten, de Gruyter, 1972.

H. Ehrig, M. Pfender and H. J. Schneider, Applications of category theory to higher dimensional formal languages, Proceedings of the Tagung über Kategorien, Mathematisches Forschungsinstitut Oberwolfach, 1973, 10-12.

S. Eilenberg, The algebraicization of mathematics, The Mathematical Sciences Essays for COSRIMS, M.I.T. Press, 1969, 153-160.

S. Eilenberg, Automata, Languages and Machines, vol. A, Academic Press, 1974.

S. Eilenberg and C. C. Elgot, Iteration and recursion, Proc. Nat. Acad. Sci. 61, 1968, 378-379.

S. Eilenberg and C. C. Elgot, Recursiveness, Academic Press, 1970.

S. Eilenberg and M. P. Schutzenberger, Rational sets in commutative monoids, J. Alg. 13, 1969, 173-191.

S. Eilenberg and J. Wright, Automata in general algebras, Inf. Cont. 11, 1967, 52-70.

C. C. Elgot, The common algebraic structure of exit-automata and machines, Computing 6, 1970, 349-370.

D. L. Elliott and R. Mullans, Linear systems over partially ordered sets, to appear.

E. Engeler, Formal Languages: Automata and Structures, Markham, 1968.

E. Engeler Ed., Symposium on Semantics of Algorithmic Languages, Lecture Notes in Mathematics 188, Springer-Verlag, 1971.

J. Engelfriet, Bottomup and topdown treetransformations: a comparison, Memorandum 19, Techniche Hogeschool Twente, Netherlands, 1971.

P. L. Falb, Infinite-dimensional filtering: the Kalman-Bucy filter in Hilbert space, Inf. Cont. 11, 1967, 102-137.

M. Fliess, Deux applications de la représentation matricielle d'une série rationnelle non commutative, J. Alg. 19, 1971, 344-353.

M. Fliess, Sur certaines familles de séries formelles, thesis, Université Paris VII, 1972.

M. Fliess, Sur la réalisation des systèmes dynamiques bilinéares, C. R. Acad. Sci. Paris, A277, 1973.

I. Flügge-Lotz, Trends in the field of automatic control in the last two decades, AIAA J., 10, 1972, 721-26.

P. Freyd, Abelian Categories, Harper and Row, 1964.

P. A. Fuhrmann, On observability and stability in infinite-dimensional linear systems, J. Opt. Th. Appl. 12, 1973, 173-181.

H. Gallaire, Decomposition of linear sequential machines II, Math. Sys. Th. 4, 1969, 168-190.

H. Gallaire and M. A. Harrison, Decomposition of linear sequential machines, Math. Sys. Th. 3, 1969, 246-287.

H. Gallaire, J. N. Gray, M. A. Harrison and G. T. Herman, Infinite linear sequential machines, J. Comp. Syst. Sci. 2, 1968, 381-419.

F. R. Gantmakher, The Theory of Matrices, 2 vols., Chelsea, 1959.

J. A. Geary and E. T. Onat, Representation of non-linear dynamical systems, to appear.

J. A. Geary and E. T. Onat, Representation of nonlinear hereditary mechanical behavior, Report ORNL-TM-4525, Oak Ridge Laboratory, June 1974.

E. G. Gilbert, Controllability and observability in multivariable systems, J. SIAM Cont. 1, 1963, 128-151.

A. Gill, Introduction to the Theory of Finite State Machines, McGraw-Hill, 1962.

A. Gill, Linear Sequential Circuits, McGraw-Hill, 1966.

S. Ginsburg, An Introduction to Mathematical Machine Theory, Addison-Wesley, 1962.

A. Ginzberg, Algebraic Theory of Automata, Academic Press, 1968.

Y. Give'on, Transparent categories and categories of transition systems, in Proceedings of the Conference on Categorical Algebra at La Jolla, Springer-Verlag, 1966, 317-332.

Y. Give'on, Categories of semimodules: the categorical structural properties of transition systems, Math. Sys. Th. 1, 1967, 67-78.

Y. Give'on, On some properties of the free monoids with applications to automata theory, J. Comp. Sys. Sci. 1, 1967, 137-154.

Y. Give'on, A Categorical Review of Algebra Automata and System Theories, Symposia Mathematica 4, Istituto Nagionale di Alta Matematica, Academic Press, 1970.

Y. Give'on and M. A. Arbib, Algebra automata II: the categorical framework for dynamic analysis, Inform. Cont. 12, 1968, 346-370.

Y. Give'on and Y. Zalcstein, Algebraic structures in linear systems theory, J. Comp. Sys. Sci. 4, 1970, 539-556.

J. A. Goguen, Bibliography of algebraic work in computer and system sciences with particular emphasis on categorical algebraic work; available from the author at 3532 Boelter Hall, Computer Science Department, UCLA, Los Angeles, California 90024, USA.

J. A. Goguen, L-fuzzy sets, J. Math. Anal. Appl. 18, 1967, 145-174.

J. A. Goguen, Categories of L-sets, Bull. Amer. Math. Soc. 75, 1969, 622-624.

J. A. Goguen, The logic of inexact concepts, Synthese 19, 1969, 325-373.

J. A. Goguen, Mathematical representation of hierarchically organized systems, Global Systems Dynamics, S. Karger, 1970, 111-129.

J. A. Goguen, Systems and minimal realization, Proceedings of the IEEE Conference on Decision and Control, 1971, 42-46.

J. A. Goguen, Minimal realization of machines in closed categories, Bull. Amer. Math. Soc. 78, 1972, 777-783.

J. A. Goguen, On homomorphisms, simulations, correctness, subroutines, and termination for programs and program schemes, Proceedings of the 13th IEEE Conference on Switching and Automata Theory, 1972, 52-60.

J. A. Goguen, Realization is universal, Math. Sys. Th. 6, 1973, 359-374.

J. A. Goguen, System theory concepts in computer science, Proceedings of the Sixth Hawaii International Conference on System Sciences, 1973, 77-80.

J. A. Goguen, On homomorphisms, correctness, termination, unfoldments and equivalence of flow diagram programs, to appear as a UCLA Computer Science Department Technical Report.

J. A. Goguen, Axioms, extensions and applications for fuzzy sets: languages and the representation of concepts, to appear.

J. A. Goguen, Some comments on applying mathematical system theory, to appear.

J. A. Goguen, J. W. Thatcher, E. G. Wagner and J. B. Wright, A junction between computer science and category theory, I: basic concepts and examples (part 1), IBM Research Report RC 4526, T. J. Watson Research Center, 1973.

J. A. Goguen and R. Yacobellis, The Myhill functor, input-reduced machines, and generalized Krohn-Rhodes theory, Proceedings of the Fifth Princeton Conference on Information Sciences and Systems, 1972, 474-478.

G. Grätzer, Universal Algebra, Van Nostrand, 1967.

J. Gray, Sheaves with values in a category, Topology 3, 1965, 1-18.

H. Halkin, An abstract framework for the theory of process optimization, Bull. Amer. Math. Soc., 1967.

M. A. Harrison, Introduction to Switching and Automata Theory, McGraw-Hill, 1965.

M. A. Harrison, Lectures on Linear Sequential Machines, Academic Press, 1969.

J. Hartmanis and R. E. Stearns, Algebraic Structure Theory of Sequential Machines, Prentice-Hall, 1966.

M. L. J. Hautus and G. H. Olsder, A Uniqueness Theorem for Linear Control Systems with Coinciding Reachable Sets, SIAM J. Cont. 11, 1973, 412-416.

A. Heller, Probabilistic automata and stochastic transformations, Math. Sys. Th. 1, 1967, 197-208.

J. Helton, Discrete time systems, operator models and scattering theory, J. Func. Anal., to appear.

J. Helton and W. Helton, Scattering theory for computers and other nonlinear systems, Proceedings of the First International Symposium: Category Theory Applied to Computation and Control, Univ. of Mass., 1974.

H. Herrlich and G. Strecker, Category Theory, Allyn and Bacon, 1973.

Carl Hewitt, ACTORS, S. F. AI conference, summer 1973.

B. L. Ho, An effective construction of realizations from input/output descriptions, dissertation, Stanford University, 1966.

B. L. Ho and R. E. Kalman, Effective construction of linear state-variable models from input/output functions, Regelungstechnik 14, 545-548.

J. E. Hopcroft and J. D. Ullman, Formal Languages and Their Relation to Automata, Addison-Wesley, 1969.

G. Hotz, Strukturelle Verwandtschaften von semi-Thue-Systemen, Proceedings of the First International Symposium: Category Theory Applied to Computation and Control, Univ. of Mass., 1974.

G. Hotz and H. Walter, Automatentheorie und Formale Sprachen II, Mannheim, 1969.

G. Hotz and V. Claus, Auotmatentheorie und Formale Sprachen III, Bibliographisches Institut Mannheim, 1971.

E. T. Irons, A syntax directed compiler for ALGOL 60, Comm. ACM 4, 1961, 51-55.

A. Isidori, Direct construction of minimal bilinear realizations from nonlinear input/output maps, IEEE Trans. Aut. Cont. AC-18, 1973, 626-631.

R. DeB. Johnson, Linear systems over various rings, M.I.T. Report ESL-R-497.

R. E. Kalman, Canonical structure of linear dynamical systems, Proc. Nat. Acad. Sci. (USA) 48, 1962, 596-600.

R. E. Kalman, A mathematical description of linear dynamical systems, J. SIAM Cont. Ser. A 1, 1963, 152-192.

R. E. Kalman, Algebraic structure of linear dynamical systems. I. The module of Σ, Proc. Nat. Acad. Sci. (USA) 54, 1965, 1503-1508.

R. E. Kalman, Algebraic aspects of the theory of dynamical syetems, in Differential Equations and Dynamical Systems, Academic Press, 1967, 133-146.

R. E. Kalman, Kronecker invariants and feedback, Proceedings of the Conference on Ordinary Differential Equations, NRL Mathematics Research Center, 1971.

R. E. Kalman, On minimal partial realizations of a linear input/output map, in Aspects of Network and System Theory, Holt, Rinehart and Winston, 1971, 385-407.

R. E. Kalman, Global structure of classes of linear dynamical systems, extended abstract of lectures presented at the NATA Advanced Study Institute on Geometric and Algebraic Methods for Nonlinear Systems, London, 1973; copies available from the author at the Center for Mathematical Systems Theory, University of Florida, Gainesville, Florida 32601, USA.

R. E. Kalman and R. S. Bucy, New results in linear prediction and filtering theory, J. Basic Engr. (Trans. ASME, Ser. D), 83D, 1961, 95-100.

R. E. Kalman, P. L. Falb and M. A. Arbib, Topics in Mathematical System Theory, McGraw-Hill, 1969.

R. E. Kalman and M.L.J. Hautus, Realization of continuous time linear dynamical systems: rigorous theory in the style of Schwartz, in Conference on Differential Equations, NRL Math Research Center, 1971, 14-23.

R. E. Kalman, Y. C. Ho and K. Narendra, Controllability of linear dynamical systems, Contr. to Diff. Eq. 1, 1963, 189-213.

E. Kamen, Control of linear continuous-time systems defined over rings of distributions, Proceedings of the First International Symposium: Category Theory Applied to Computation and Control, Univ. of Mass., 1974.

E. Kamen, Algebraic results on time-varying systems, to appear.

E. Kamen, On an algebraic theory of systems defined by convolution operators, to appear.

E. Kamen, Topological module structure of linear continuous-time systems, SIAM J. Cont., to appear.

W. H. Kautz ed., Linear Sequential Switching Circuits - Selected Technical Papers, Holden-Day, 1965.

Ikuo Kimura, Categories of Local Dynamical Systems, Funkcialaj Ekvakioj, 16, 1973, 29-52.

A. Kock, Monads on symmetric monoidal closed categories, Arch. Math. XXI, 1970, 1-10.

P. Kossowski, Functorial formulations of models of first order theories, Masters Thesis, Univ. of Ottawa, in preparation.

K. B. Krohn and J. L. Rhodes, Algebraic theory of machines. I. The main decomposition theorem. Trans. Amer. Math. Soc. 116, 1965, 450-464.

V. Kučera, Algebraic theory of discrete optimal control for single-variable systems II, III, Kybernetika 9, 1973, 206-221 and 291-312.

J. Lambek, Deductive systems and categories, Math. Sys. Th. 2, 1968, 287-318.

P. J. Landin, A program machine symmetric automata theory, Mach. Intell. 5, 1970.

F. W. Lawvere, Functorial semantics of algebraic theories, dissertation, Columbia University, 1963.

P. D. Lax and R. S. Phillips, Scattering Theory, Academic Press, 1967.

C. F. Lee, Control systems design and optimization, a rigorous mathematical approach by means of a mathematical theory of systems engineering, master's thesis, Univ. Arizona, 1973.

L. S. Levy and A. K. Joshi, Some results in tree automata, Math. Sys. Th. 6, 1973, 334-342.

F.E.J. Linton, Relative functorial semantics: adjointness results, Lecture Notes in Mathematics 99, Springer-Verlag, 1969, 384-418.

D. G. Luenberger, Observing the state of a linear system, IEEE Trans. Military Electronics, MIL-8, 1964, 74-80.

D. G. Luenberger, An introduction to observers, IEEE Trans. Aut. Cont. AC-16, 1971, 596-602.

S. Mac Lane, Categories for the Working Mathematician, Springer-Verlag, 1972.

E. G. Manes, Algebraic Theories, Springer-Verlag, to appear.

Z. Manna and J. Vuillemin, Fixpoint approach to the theory of computation, Comm. ACM 15, 1972, 528-536.

W. S. McCulloch and W. H. Pitts, A logical calculus of the ideas immanent in nervous activity, Bull. Math. Biophys. 5, 1943, 115-33.

M. D. Mesarovic, Auxiliary functions and constructive specification of general systems, J. Math. Sys. Th. 2, 1968.

M. D. Mesarovic, On some metamathematical results as properties of general systems, J. Math. Sys. Th. 2, 1968.

M. D. Mesarovic, Controllability of general systems, J. Math. Sys. Th. 5, 1971.

M. D. Mesarovic and Y. Takahara, Foundations for a mathematical systems theory, to appear.

W. Merzenich, Cellular automata with additive local transition, Proceedings of the First International Symposium, Category Theory Applied to Computation and Control, Univ. of Mass., 1974.

J. Meseguer and I. Sols, Categorical tensor representation of deterministic, relational and probabilistic finite functions, Actas de las II Jornadas Matemáticas Hispano-Lusitanas, 1973, to appear.

J. Meseguer and I. Sols, Automata in semimodule categories, Proceedings of the First International Symposium: Category Theory Applied to Computation and Control, Univ. of Mass., 1974.

J. Mezei and J. B. Wright, Algebraic automata and context-free sets, Inf. Cont. 11, 1967, 3-29.

J. Mezei and J. B. Wright, Generalized ALGOL-like languages, Inf. Cont. 11, 1967.

B. Mitchell, Theory of Categories, Academic Press, 1965.

E. G. Moore, Gedanken experiments on sequential machines, Automata Studies, Princeton Univ. Press, 1956, 129-153.

M. Morcrette, On a category of semi automata coming from the theory of equations in the free monoid, to appear.

A. S. Morse, Geometric concepts in linear system theory, to appear.

A. S. Morse and W. M. Wonham, Status of noninteracting control, IEEE Trans. Aut. Cont. AC-16, 1971, 568-581.

J. Myhill, Finite automata and the representation of events, WADC Technical Report Wright-Patterson AFB, 1957.

A. Nerode, Linear automaton transformations, Proc. Amer. Math. Soc. 9, 1958, 541-544.

A. Nerode, Some Stone spaces and recursion theory, Duke Math. J. 26, 1959, 397-406.

M. Nivat, Transductions des langages de Chomsky, Annales Institut Fourier (Grenoble) 18, 1968, 339-456.

Janusz Nowakowski, Studies on the possibility of applying the method of classification analysis to the identification of objects, Prace Inst. Automat. PAN Zeszyt 78, 1969, 3-26. (Polish, English summary)

E. T. Onat and F. Fardshisheh, Representation of elastoplastic behavior by means of state variables, Proc. Int. Symp. on Plasticity, Warsaw, 1972, to appear.

E. T. Onat and J. A. Geary, Representation of a class of nonlinear systems, Proceedings of the First International Symposium: Category Theory Applied to Computation and Control, Univ. of Mass., 1974.

L. Padulo and M. A. Arbib, System Theory: a Unified Approach to Discrete and Continuous Systems, W. B. Saunders, 1974.

B. Pareigis, Categories and Functors, Academic Press, 1970.

A. Paz, Introduction to Probabilistic Automata, Academic Press, 1971.

J. M. Perry and N. E. Sondak, A categorical approach to programs and program computation, to appear.

L. Petrone, Syntax directed mapping of context-free languages, Proceedings of the Ninth Annual Symposium on Switching and Automata Theory, 1968.

R. S. Pierce, Introduction to the Theory of Abstract Algebras, Holt, Rinehart and Winston, 1968.

R. S. Pierce, Classification problems, Math. Sys. Th. 4, 1970, 65-80.

V. M. Popov, Hyperstability and optimality of automatic systems with several control functions, Rev. Roum. Sci. Tech. - Electrotechn. et Energ. 9, 1964, 629-690.

V. M. Popov, Invariant description of linear, time-invariant controllable systems, SIAM J. Cont. 10, 1972, 252-264.

R. E. Prather, On categories of infinite automata, Math. Sys. Th. 4, 1970, 295-305.

M. O. Rabin and D. Scott, Finite automata and their decision problems, IBM J. Res. Dev. 3, 1959, 114-125.

G. Raney, Sequential functions, J. Ass. Comp. Mach. 5, 1958, 177-180.

D. C. Rine, A theory of general machines and functor projectivity, dissertation, Univ. of Iowa, 1970; available as number 70-23939 from Univ. Microfilms, Ann Arbor, Michigan, USA.

J. Rissanen and T. Kailath, Partial realization of random systems, Proceedings of the 5th IFAC Conference, 1972.

J. Rissanen and B. F. Wyman, Duality of input/output maps, Proceedings of the First International Symposium: Category Theory Applied to Computation and Control, Univ. of Mass., 1974.

S. Rolewicz, On optimal observability of linear systems with infinite-dimensional states, Stud. Math. 44, 1972, 411-416.

R. Rosen, Abstract biological systems as sequential machines, I., Bull. Math. Biophys. 26, 1964, 103-111.

R. Rosen, Abstract biological systems as sequential machines, III, Bull. Math. Biophys. 28, 1966, 141-148.

H. H. Rosenbrock, Multivariable and State-Space Theory, Wiley, 1970.

Y. Roucheleau, Finite-dimensional, constant, discrete-time linear dynamical systems over some classes of commutative rings, dissertation, Stanford University, 1972.

Y. Roucheleau and B. F. Wyman, Linear dynamical systems over integral domains, J. Comp. Sys. Sci., to appear.

Y. Roucheleau, B. F. Wyman and R. E. Kalman, Algebraic structure of linear dynamical systems III. realization theory over a commutative ring, Proc. Nat. Acad. Sci. (USA) 69, 1972, 3404-3406.

W. C. Rounds, Mappings and grammars on trees, Math. Sys. Th. 4, 1970.

C.-P. Schnorr, Transformational classes of grammars, Inf. Cont. 14, 1969, 252-277.

D. Scott, Some definitional suggestions for automata theory, J. Comp. Sys. Sci. 1, 1967, 187-212.

D. Scott, The lattice of flow diagrams, Symposium on Semantics of Algorithmic Languages, Lecture Notes in Mathematics 188, Springer-Verlag, 1971.

D. Scott, Continuous lattices, Proceedings of the 1971 Dalhousie Conference, Lecture Notes in Mathematics 274, Springer-Verlag, 1972, 97-136.

D. Scott, Latice-theoretic models for the λ-calculus, to appear.

D. Scott and C. Strachey, Toward a mathematical semantics for computer languages, Proceedings of the Symposium on Computers and Automata, Microwave Research Institute 21, Polytechnic Institute of Brooklyn, 1971.

Z. Semadeni, The category of logical kits, Proceedings of the Tagung über Kategorien, Mathematisches Forschungsinstitut Oberwolfach, 1972, 50-51.

C. E. Shannon, A mathematical theory of communication, Bell Sys. Tech. J. 27, 1948, 379-423 and 623-656.

C. D. Shepard, Languages in general algebras, Proceedings of the ACM Symposium on the Theory of Computing, 1969, 155-163.

L. M. Silverman, Realization of linear dynamical systems, IEEE Trans. Aut. Cont. AC-16, 1971, 554-567.

L. M. Silverman and B.D.O. Anderson, Controllability, observability and stability of linear systems, SIAM J. Contr. 6, 1968, 121-130.

T. L. Sipusic, Two automata decomposition theorems generalized to machines in a closed monoidal category, Institute for Computer Research, Univ. of Chicago, Quarterly Report No. 35, 1972.

C. V. Srinivasan, State diagram of linear sequential machines, J. Franklin Inst. 273, 1962, 383-418.

P. H. Starke, Abstrakte Automaten, VEB-Verlag, Berlin 1969; English translation: Abstract Automata, Elsevier, North Holland, 1972.

P. H. Starke, Allgemeine Probleme und Methoden in der Automatentheorie, EIK 8, 1972, 489-517.

M. Steinby, On pair algebras and state-information in automata, Ann. Acad. Sci. Fennicae 542, 1973.

Noboru Sugie, Reducible linear time-varying control systems, International J. Control 14, 1971, 149-160.

Y. Takahara and M. D. Mesarovic, Application of categorical algebra to classification of systems, to appear.

J. W. Thatcher, Characterizing derivation trees of context-free grammars through a generalization of finite automata theory, J. Comp. Sys. Sci. 1, 1967.

J. W. Thatcher, There's a lot more to finite automata theory than you would have thought, IBM T. J. Watson Research Center Report RC 2852, 1970.

J. W. Thatcher, Generalized^2 sequential machine maps, J. Comp. Syst. Sci. 4, 1970, 339-367.

J. W. Thatcher and J. B. Wright, Generalized finite automata theory, Math. Syst. Th. 2, 1968, 57-81.

W. Vollmerhaus, Über die Zerlegung von frein X-Kategorien, in 4. Colloquium uber Automatentheorie, TU München, 1967.

E. Wagner, An algebraic theory of recursive definitions and recursive languages, Proceedings of the third annual ACM Symposium on the Theory of Computing, 1971.

E. Wagner, Languages for defining sets in arbitrary algebras, Proceedings of the Twelfth IEEE Symposium on Switching and Automata Theory, 1971.

H. Walter, Verallgemeinerte Pullbackkonstruktionen bei Semi-Thue-Systemen und Grammatiken, EIK 6, 1970, 239-254.

H. Walter, Einige topologische Aspekte der syntaktischen Analyse und Ubersetzung bei Chomsky-Grammatiken, J. Comp. Sys. Sci., to appear.

M. Wand, An unusual application of program-proving, Proceedings of the Fifth ACM Symposium on the Theory of Computing, 1973, 59-66.

M. Wand, Mathematical foundations of language theory, dissertation, Project MAC, M.I.T., 1973.

M. Wand, An algebraic formulation of the Chomsky hierarchy, Proceedings of the First International Symposium: Category Theory Applied to Computation and Control, Univ. of Mass., 1974.

M. Wand, On the recursive specification of data types, Proceedings of the First International Symposium: Category Theory Applied to Computation and Control, Univ. of Mass., 1974.

M. Wand, On the behavioral description of data structures, to appear.

M. E. Warren and A. E. Eckberg, On the dimensions of controllability subspaces: a characterization via polynomial matrices and Kronecker invariants, SIAM J. Cont., to appear.

L. Weiss and R. E. Kalman, Contributions to linear system theory, Intern. J. Engr. Sci. 3, 1965, 141-171.

J. C. Willems and S. K. Mitter, Controllability, observability, pole allocation and state reconstruction, IEEE Trans. Aut. Cont. AC-16, 1971, 582-595.

T. G. Windeknecht, Mathematical systems theory: causality, Math. Sys. Th. 1, 1967.

A. Wiweger, On coproducts of automata, Bull. Acad. Polonaise Sci. 21, 1973, 753-758.

B. F. Wyman, Linear systems over rings of operators, Proceedings of the First International Symposium: Category Theory Applied to Computation and Control, Univ. of Mass., 1974.

A. W. Wymore, A wattled theory of systems, in Trends in General System Theory, Wiley, 1972.

A. W. Wymore, Discrete systems and continuous systems, in Advances in Mathematical Systems Theory, Penn. State Univ. Press.

A. W. Wymore, The tricotyledon theory of system design, Proceedings of the First International Symposium: Category Theory Applied to Computation and Control, Univ. of Mass., 1974.

A. W. Wymore, A Mathematical Theory of Systems Engineering: The Elements, Wiley, 1967.

S. S. Yau and K. C. Wang, Linearity of sequential machines, IEEE Trans. Elec. Comp. EC-15, 1966, 337-354.

L. Zadeh and C. A. Desoer, Linear Systems Theory, McGraw-Hill, 1963.

H. P. Zeiger, Ho's algorithm, commutative diagrams, and the uniqueness of minimal linear systems, Inf. Cont. 11, 1967, 71-79.

H. N. Zessin, Kategorientheoretische Behandlung der Zustandsraumtransformation von Markoffprozessen, J. Reine Ang. Math., 260, 1973.

SUPPLEMENTARY BIBLIOGRAPHY

B. A. Asner and A. Halanay, Algebraic theory of pointwise degenerate delay-differential systems, J. Diff. Eq. 14, 1973, 293-306.

F. S. de Blasi and J. Schinas, Stability of multivalued discrete dynamical systems, J. Diff. Eq. 14, 1973, 245-262.

H. Ehrig, M. Pfender and H.-J. Schneider, Graph-grammars, an algebraic approach, Proc. Conf. IEEE SWAT, 1973.

H.-J. Hoehnke, Synthesis and complexity of logical systems, to appear.

C. A. R. Hoare, An axiomatic basis for computer programming, Comm. ACM 12, 1969.

P. J. Landin and R. M. Burstall, Programs and their proofs: an algebraic approach, Machine Intelligence 4, American Elsevier, 1969.

J. McCarthy, Towards a mathematical theory of computation, Proc. IFIP Congress 1962, North Holland, 1963.

M. S. Paterson, Program Schemata, Machine Intelligence 3, American Elsevier, 1968.

J. Riguet, Sur les rapports entre les concepts de machine de multipole et de structure algebrique, C. R. Acad. Sci. Paris, t 237, 1953, 425-427.

J. Riguet, Programmation et theorie des categories, Symposium on Symbolic Languages in Data Processing, Rome, Gordon and Breach, 1962, 83-98.

A. S. Willsky, Invertibility of finite group homomorphic sequential systems, to appear.

R. H. Yacobellis, Toward a generalized Krohn-Rhodes theory, to appear.